BEHAVIORAL
NEUROCHEMISTRY

About the editors . . .

José M. R. Delgado received his M.D. from Madrid
University in 1940 and his D.Sc. in Neurophysiology from Madrid University in 1942. He taught
Physiology and Psychiatry at Yale University
Medical School from 1952 to 1974. Since 1974,
he has been Professor and Chairman of the Depart-
ment of Physiological Sciences at the Autonomous
University Medical School, Madrid, and Director
of Research at the National Institute "Ramón y
Cajal," Madrid. He has received international
acclaim for his work in Neurophysiology, receiving
the Rodriguez Pascual Prize in 1975, the Gold
Medal Award of the Society of Biological Psychiatry
in 1974, the Gold Medal Exhibit Award of the
Americal Psychiatric Society, a Guggenheim
Fellowship in 1963, the Spanish government's
Ramón y Cajal Prize in 1952, the Roel Prize in
1945, and the Countess of Maudes Prize in 1944.

Francis V. DeFeudis received his B.S. in Zoology
from Boston University and the University of
Miami in 1961, his M.A. in Biology from Clark
Univeristy, Worcester, Massachusetts, in 1964,
and his Ph.D. in Biochemistry from McGill Uni-
versity, Montreal, in 1968. He held Post-Doctoral
Fellowships in Pharmacology at McGill University
(1968) and in Psychiatry at Yale University Medical
School (1968-69). He held Wellcome Research
Fellowships in Anaesthesia Research at McGill
University (1969-70) and Pharmacology at the
University of Cambridge, England (1970). He
has been Head of the Section of Neurochemistry
at the National Institute "Ramón y Cajal" since
1974 and is Professor of Physiology at the
Autonomous University Medical School, both in
Madrid.

BEHAVIORAL NEUROCHEMISTRY

Edited by

J.M.R. Delgado, M.D.

and

F.V. DeFeudis, Ph.D.

Both of the Department of Physical Sciences
Autonomous University Medical School
Madrid, Spain

SP

S P Books Division of
SPECTRUM PUBLICATIONS, INC.
New York

Distributed by Halsted Press
A Division of John Wiley & Sons

New York   Toronto   London   Sydney

SPECTRUM PUBLICATIONS, INC.
86-19 Sancho Street, Holliswood, N.Y. 11423

Distributed solely by the Halsted Press division of John Wiley & Sons, Inc., New York

Library of Congress Cataloging in Publication Data

Main entry under title:

Behavioral neurochemistry.

   1. Psychology, Physiological--Congresses.
2. Neurochemistry--Congresses. 3. Neuropsychophar-
macology--Congresses. 4. Animals, Habits and behavior of--Congresses. I. Delgado, José Manuel R.,
1915-   II. DeFeudis, F. V.
QP360.B43   152   76-22627
ISBN 0-470-15179-X

Contents

Preface

1

New Trends in Behavioral Neurochemistry 1
   J. M. R. Delgado

2

New Trends in Behavioral Neurochemistry 11
   F. V. DeFeudis

3

Amine Regulation of Protein Synthesis in Retrograde Amnesia 25
   W. B. Essman and Shirley G. Essman

4

Neurochemical Correlates of Learning Ability 63
   Y. Tsukada, M. Nomura, K. Nagai, S. Kohsaka,
   H. Kawahata, M. Ito and T. Matsutani

CONTENTS

5

Alterations in the Structure and Function of Free and Membrane-Bound Polysomes and Messenger-RNA of Neuronal and Glial Enriched Fractions of Rat Cerebral Cortex During Light Deprivation and Exposure to Lights of Different Wavelengths 85
M. R. V. Murthy, Huguette Roux, A. D. Bharucha and R. Charbonneau

6

About a "Specific" Neurochemistry of Aggressive Behavior .. 113
L. Valzelli

7

Relationships among Rhythmic Slow Waves in the Brainstem, Monoamines, and Behavior 133
Takashi Tsubokawa

8

Influence of Some Mediators upon the Metabolic Activity of Neuronal Subcellular Structures 155
Nina Docheva Georgieva and Radoy Ivanov Ivanov

9

Effects of Ethanol on Central Nervous System Cells 165
Ernest P. Noble

10

Cerebellar Cyclic Nucleotides, Prostaglandin E_2 and Some Convulsant and Tremorogenic Drugs 183
R. Fumagalli, F. Berti, G. C. Folco, D. Longiave and R. Paoletti

11

Recognins and Their Chemoreciprocals 197
Samuel Bogoch

12

A Neurobiological Theory of Action of Lithium in the Treatment of Manic-Depressive Psychosis 223
Arnold J. Mandell and Suzanne Knapp

Index 251

Preface

One of the most challenging biological problems of our time concerns the elucidation of neurochemical events' underlying behavior. Therefore, as part of the Fifth International Meeting of the International Society for Neurochemistry, we organized a satellite symposium on Behavioral Neurochemistry, held at the new Autonomous Medical School in Madrid, which dealt with this subject in some detail. Speakers from laboratories in ten countries presented their recent work on environmental, genetic, and maturational effects on CNS biochemistry, physiology, and behavior. This book makes available expanded versions of the presentations of contributors to this meeting as well as the work of several investigators who were unable to participate. Hopefully, its contents will serve to stimulate further development of ideas about brain chemistry and behavior.

Thanks are due to the speakers and to other participants of this meeting, to Mrs. Caroline S. Delgado for her role in organizing the meeting, and to Spectrum Publications, Inc., for their interest in publishing its proceedings.

December, 1975

J. M. R. DELGADO, M.D.
F. V. DE FEUDIS, Ph.D.
Madrid, Spain

Contributors

F. Berti, M.D.
Institute of Pharmacology and
    Pharmacognosy
University of Milan
Milan, Italy

A.D. Bharucha, Ph.D.
Department of Biochemistry
Faculty of Medicine
Laval University
Quebec, Canada

S. Bogoch, M.D., Ph.D.
Foundation for Research on the Nervous
    System
Boston University School of Medicine, and
The Dreyfus Medical Foundation
New York, New York
U.S.A.

R. Charbonneau, Ph.D.
Department of Biochemistry
Faculty of Medicine
Laval University
Quebec, Canada

F.V. DeFeudis, Ph.D.
National Center "Ramón y Cajal," and
Autonomous University Medical School
Madrid, Spain

Shirley G. Essman, Ph.D.
Department of Philosophy
State University of New York at
    Stony Brook
Stony Brook, New York
U.S.A.

Walter B. Essman, M.D., Ph.D.
Departments of Psychology and
    Biochemistry
Queens College of the City University of
    New York
Flushing, New York
U.S.A.

G.C. Folco, M.D.
Institute of Pharmacology and
    Pharmacognosy
University of Milan
Milan, Italy

CONTRIBUTORS

R. Fumagalli, M.D.
Institute of Pharmacology and
    Pharmacognosy
University of Milan
Milan, Italy

Nina Docheva Georgieva, Dr. Biol.
Department of Human and Animal
    Physiology
Faculty of Biology
University of Sofia
Sofia, Bulgaria

Masato Ito, M.A.
Department of Physiology
Keio University School of Medicine
Tokyo, Japan

Radoy Ivanov Ivanov, Ph.D.
Department of Human and Animal
    Physiology
Faculty of Biology
University of Sofia
Sofia, Bulgaria

Hisako Kawahata, B.S.
Department of Physiology
Keio University School of Medicine
Tokyo, Japan

Suzanne Knapp, B.A.
Department of Psychiatry
School of Medicine
University of California at San Diego
La Jolla, California
U.S.A.

Shinichi Kohsaka, M.D.
Department of Physiology
Keio University School of Medicine
Tokyo, Japan

D. Longiave, M.D.
Institute of Pharmacology and
    Pharmacognosy
University of Milan
Milan, Italy

Arnold J. Mandell, M.D.
Departments of Psychiatry, Neurosciences,
    Physiology, and Pharmacology
School of Medicine
University of California at San Diego
La Jolla, California
U.S.A.

Tenhoshimaru Matsutani, M.D.
Department of Developmental Physiology
School of Medicine
Fujita-Gakuen University
Kutsukake-cho Toyoake City
Japan

M.R.V. Murthy, **Ph.D.**
Department of Biochemistry
Faculty of Medicine
Laval University
Quebec, Canada

Katsuko Nagai, B.S.
Department of Physiology
Keio University School of Medicine
Tokyo, Japan

E.P. Noble, Ph.D., M.D.
National Institute on Alcohol Abuse and
    Alcoholism
Rockville, Maryland
U.S.A.

Masahiko Nomura, M.D.
Department of Physiology
Keio University School of Medicine
Tokyo, Japan

R. Paoletti, M.D.
Institute of Pharmacology and
    Pharmacognosy
University of Milan
Milan, Italy

H. Roux, **Ph.D.**
Department of Biochemistry
Faculty of Medicine
Laval University
Quebec, Canada

Takashi Tsubokawa, M.D., M.D.Sc.
Department of Neurological Surgery
School of Medicine
Nihon University
Tokyo, Japan

Yasuzo Tsukada, M.D.
Department of Physiology
Keio University School of Medicine
Tokyo, Japan

L. Valzelli, M.D.
Istituto di Ricerche Farmacologiche
  "Mario Negri"
Milan, Italy

New Trends in Behavioral Neurochemistry

J. M. R. DELGADO

National Center "Ramón y Cajal"
and Autonomous University
Medical School
Madrid, Spain

Current researches into the neurochemical bases of behavior follow two divergent but closely related trends. At one end of the spectrum are studies of subcellular particles and unitary activity, and at the other, studies of whole organisms and their social groups. The scope of this volume illustrates some of the complementary aspects of neurochemical investigations, as the contributions range from the *in vitro* analysis of mitochondrial suspensions (Georgieva and Ivanov) to an examination of the neurochemistry of aggressive behavior (Valzelli).

A. PUSH-PULL CANNULAS, CHEMITRODES, AND DIALYTRODES

Neurochemical investigations in general require killing of the experimental animal, removal of the brain, and homogenization of the tissue in order to analyze and quantitate the contents and activities of possible transmitters, enzymes, and other substances. Use of these procedures precludes repetition of experiments in the same subject, and the significance of data must be established by statistical evaluation of pooled data from groups of animals. Adequate control requires the sacrifice of a similar group of animals with subsequent analysis of their brains. Instead of stopping the ongoing cerebral metabolic activity at a determined moment of the animal's behavior history, a new research trend aims at testing repeatedly in the same subject the correlations between chemical processes and behavioral manifestations. For this purpose, several workers have proposed the use of devices which allow intracerebral perfusion of discrete cerebral structures in fully awake animals.

A *push-pull cannula* (Gaddum, 1961) consists of two tubings, juxtaposed or concentric, is usually implanted stereotaxically in the brain, and allows the simultaneous injection and collection of fluids. This device permits the perfusion of discrete cerebral structures, or in some cases, the cerebral ventricles (Pappenheimer et al., 1962).

A *chemitrode* (Delgado et al., 1962) is a push-pull cannula plus an array of electrodes placed alongside the tubings which permits chronic repetitions of the following procedures in the awake animal: (a) electrical recording; (b) electrical stimulation; (c) injection of substances; (d) collection of intracerebral fluids; and (e) perfusion of liquids through discrete areas of the brain.

A *dialytrode* (Delgado, 1971; Delgado et al., 1972) is a chemitrode with its tip connected to a small semipermeable bag which serves as a barrier for microorganisms and tissue cells, while permitting the passage of fluids and chemicals.

A *transdermal dialytrode* (Delgado et al., 1972) is a totally implantable device, with two subcutaneous reservoirs that permits transdermal injection and collection of fluids to and from the brain, as well as electrical stimulation and recording. This instrument was designed for possible therapeutic application in man.

The main advantages of the dialytrode system are that it (a) minimizes the risk of infection; (b) prevents blocking of cannulas; (c) avoids pressure in the brain; and (d) assures good recovery of perfusates. The main disadvantage of the system is that the presence of a membrane slows down the transfer of chemicals between the perfusion fluid and the brain. *In vitro* studies show that by immersing a dialytrode in a small bath and circulating synthetic spinal fluid through the bag at a rate of 4 μl/min, labeled compounds such as ^{3}H-tyrosine passed the membrane at a speed of -0.5% of total tyrosine/hr, which is a substantial amount considering that the capacity of the dialytrode bag is only 3 μl. *In vivo* studies in the monkey with dialytrodes implanted in the amygdala show further that perfusion of L-glutamate (1 M) at a rate of 4 μl/min produced typical glutamate seizure in about 10 min, and this abnormal activity lasted for 2–6 min.

Following perfusion of synthetic spinal fluid through the amygdala, the collected perfusate contained glutamine, asparagine, serine, glutamate, glycine, and α-alanine, thus demonstrating the spontaneous release of these substances from the amygdala. In other experiments, U^{14}-C-D-glucose was added to ringer fluid circulated through the amygdala at a rate of 1.2 μl/min for 1 hr. After a wash, Ringer solution was perfused at the same rate for 12 hr. Analyses indicated the presence on the perfusate of labeled aspartate and citrulline, a finding which indicated that (a) labeled glucose had diffused to the brain; (b) the new amino acids had been synthesized in the amygdala; and (c) the labeled amino acids had diffused back to the dialytrode bag.

Push-pull cannulas and chemitrodes have been widely used in the last decade, representing an important methodological advance in neurochemical and neuropharmacological studies of behavior (e.g., Chase and Kopin, 1968; Sparber, 1975; Stein and Wise, 1969; Winston and Gerlach, 1971; see reviews by Delgado et al., 1972, and Myers, 1974).

Using the chemitrode technology in monkeys, we have shown that dopamine and noradrenaline can be synthesized locally in the caudate nucleus, hypothalamus, and amygdala from their precursors tyrosine and dopa (Roth et al., 1969). Tyrosine injected into the caudate nucleus was preferentially metabolized to dopamine with the mean noradrenaline/dopamine proportion of 1/61. In contrast, the hypothalamus and amygdala converted tyrosine preferentially into noradrenaline. When dopa was used as a precursor, the ratio of noradrenaline/dopamine was about 1/13 and was similar in the three brain structures tested. These findings indicated the considerable microanatomic differences that exist in the distribution of enzymes which regulate the turnover of tyrosine, whereas the distribution of amino acid decarboxylase activity was nonspecific.

B. REGIONAL NEUROCHEMISTRY

Analyses of brain homogenates have provided fruitful results, but these studies have obvious limitations because discrete, local changes in chemistry are undetectable when large amounts of brain are pooled together. For example, it is well known that the locus coeruleus and raphé nuclei play very different roles in the synchronization of sleep and have diverse pharmacological sensitivity and metabolic activity. However, a pooled sample of mesencephalic tissue would not give this information. Also, we should consider that most cerebral structures (e.g., caudate nucleus or hypothalamus) have great behavioral, physiological, and chemical complexity, since they are closely related to other structures. For this reason, it would be preferable to study the altered state of equilibrium among several structures and substances, rather than to evaluate the possible changes of one transmitter in one area. One of the advantages of using chemitrodes is that several cerebral areas may be perfused, permitting the study of chemical changes that take place simultaneously in different structures.

With the standard procedure of killing the experimental animal and removing its brain, regional neurochemsitry is usually based on the selection of a specific structure which is dissected away from the rest of the brain. However, the anatomic name of an area does not imply functional or chemical homogeneity. A recent research trend involves the further subdivision of brain regions. An example is provided by the following preliminary results obtained in our department (Echandía et al., in preparation). Rabbits were implanted with electrodes in the hippocampus, and one side was electrically stimulated with 125 μA for 1 hr daily for 10 days in order to investigate the possible enduring chemical changes induced by this repeated excitation. Analyses of the right and left hippocampi did not reveal any significant changes in the activities of acetylcholinesterase and acid phosphatase. The experiments were repeated in another group of rabbits, but this time the analyses were performed after dividing the hippocampus into six different regions. As shown in Fig. 1, the activities of both cholinesterase and acid phosphatase were significantly increased in the stimulated animal, and this change was restricted to a discrete region of the dorsal hippocampus without being manifest in other parts. These results indicated a regional selectivity of neurochemical changes which would have been undetectable if the whole hippocampus had been analyzed.

C. THE NEUROCHEMISTRY OF INDIVIDUALITY

Just as pooling large areas of the brain may obscure metabolic changes that occur in local cerebral regions, the usual pooling of chemical and behavioral data in order to evaluate their statistical significance may conceal the special reactivity of experimental subjects. For this reason, among others, studies on the chemical and neurophysiological bases of individuality are usually neglected.

It is well known, however, that behavioral responses are determined by a combination of genetic factors, environmental circumstances, and

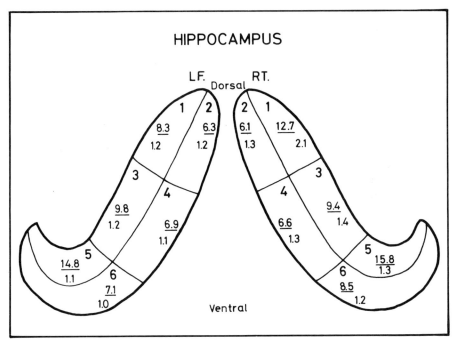

1 Diagram of the left (LF) and right (RT) hippocampi of the rabbit. This structure was dissected out of the brain and divided in six sectors, as shown by numbers 1 to 6. Each part was analyzed for its acetylcholinesterase activity (μmoles/mg/min), expressed by numbers underlined, and its acid phosphatase activity (μmoles/gr/min), expressed by lower numbers *not* underlined. Electrical stimulations were applied to sector 1 of the right hippocampus. The acetylcholinesterase activity of sector 1 was significantly increased (12.7) with respect to both the contralateral side (8.3) and other sectors of the hippocampus. Acid phosphatase was also increased in sector 1 (1.2).

individual experiences, and that all these elements may differ from subject to subject. The deterrent in the investigation of individuality is the complexity and lack of knowledge about the many elements involved. In spite of these difficulties, with suitable experimental design some aspects may be investigated, and this trend could be important for behavioral neurochemistry.

As a model, we have proposed the study in monkeys of pharmacological reactivity of the same subject while alone, or when paired in a dominant or submissive situation (Delgado et al., 1976). Experimental

2 Studies of "free" behavior are possible by placing animals in a cage. Recording of data and delivery of electrical and/or chemical intracerebral stimuli can be performed by remote control.

results have demonstrated that some behavioral categories were modified significantly depending on the social status of the animal. It was also shown that the monkey had much greater sensitivity to diazepam when it was submissive than when it was dominant. The finding that individual rank is an important determinant of the pharmacological effects of a tranquilizing agent paved the way for further testing of psychoactive drugs and for the study of neurochemical mechanisms involved in the maintenance of social hierarchy and of individual characteristics of behavior. In the literature and in this volume, there exists abundant information which indicates that individuals possess different bio-chemical and neuroanatomic characteristics, depending on the level of sensory deprivation or enrichment of their environments (e.g., DeFeudis, 1975; Rosenzweig et al., 1968).

D. INSTRUMENTAL AND SPONTANEOUS BEHAVIOR

Studies on behavioral neurochemistry depend to a great extent upon the methodology and procedures used for the identification, record-ing, and quantification of the behavior. In spite of many difficulties (Dews, 1955), the fact that "... behavior is a phenomenon amenable to study by the methods of Natural Science ..." (Sidman, 1959) has been slowly recognized in psychology (Skinner and Heron, 1937), in pharmacology (Sidman, 1959), and in neurochemistry (this volume).

The development of operant behavior techniques (Skinner, 1938) provided animals with levers, compartments, mazes, and other apparati which have been the backbone for the formal quantification of behavioral responses. Manipulation of instruments represents, however, an artificiality that is rather different from the normal activities of animals. Research based on operant behavior, especially in primates, has required selection of candidates and rejection of subjects that were "lazy," "stupid," or "too destructive." Studies on the biological mechanisms and neurochemical correlates of animals that did not conform to the experimenter's qualifications could have been as interesting as research on subjects that were cooperative enough to manipulate instruments. The study of "free" behavior is an essential counterpart of instrumental studies and is deserving of greater attention in future research.

Two new trends should be mentioned in this respect: (a) the study of social behavior of free animals paired or forming a group inside a cage in the laboratory (Delgado, 1964; Delgado et al., 1973, 1976; see Fig. 2); and (b) field studies of animals completely free in a controlled environment, for example, on the island of Hall (Bermuda) where a colony of gibbons has been investigated for several years (Fig. 3;

3 The small island of Hall in Bermuda offers a controlled environment for field studies, in which telemetric and radio stimulation contact can be maintained with completely free animals.

Delgado et al., 1975). In both cases, social behavior must be quantified, and recordings of biological data as well as delivery of electrical or chemical stimulation must be performed by telemetry and by radio to avoid interference with spontaneons activities.

The study of electrophysiological and neurochemical mechanisms of the awake brain should establish links between research in different fields including subcellular particles, unitary activity, neurotransmitter agents, spontaneous behavior, and ethological observations.

E. REFERENCES

Chase, T.N., and Kopin, I.J. (1968). Stimulus-induced release of substances from olfactory bulb using the push-pull cannula. *Nature* 217, 466–467.

ˈ DeFeudis, F.V. (1975). Cerebral biochemical and pharmacological changes in differentially-housed mice. In: *Current Developments in Psychopharmacology* (W.B. Essman and L. Valzelli, eds.), pp. 143–201, Spectrum Pub., New York.

ˤ Delgado, J.M.R. (1964). Free behavior and brain stimulation. In: *International Review of Neurobiology, Vol. VI,* (C.C. Pfeiffer and J.R., Smythies, eds.), pp. 349–449, Academic Press, New York.

Delgado, J.M.R. (1971). Dialysis intercerebral. In: *Homenaje al Prof. B. Lorenzo Velazquez,* pp. 379–389, Oteo, Madrid.

Delgado, J.M.R., DeFeudis, F.V., Roth, R.H., Ryugo, D.K., and Mitruka, B.M. (1972). Dialytrode for long term intracerebral perfusion in awake monkeys. *Arch. Int. Pharmacodyn.* 198, 9–21.

Delgado, J.M.R., Grau, C., Delgado-García, J.M., and Rodero, J.M. (1976). Effects of diazepam related to social hierarchy in rhesus monkeys. *Neuropharmacology.* In press.

Delgado, J.M.R., Lipponen, V., Weiss, G., Del Pozo, F., Monteagudo, J.L., and McMahon, R. (1975). Two way transdermal communication with the brain. *Amer. Psychologist* 30, 265–273.

Delgado, J.M.R., Sanguinetti, A.M., and Mora, F. (1973). Aggressive behavior in gibbons modified by caudate and central gray stimulation. *Int. Res. Commun. System, Sept.,* 16, 2–32.

Delgado, J.M.R., Simhadri, P., and Apelbaum, J. (1962). Chronic implantation of chemitrodes in the monkey brain. *Proc. Int. Union Physiol. Sci.* 2, 1090.

Dews, P.B. (1955). Differential sensitivity to pentobarbital of pecking performance in pigeons depending on the schedule of reward. *J. Pharmacol. Exp. Ther.* 113, 393–401.

Echandia, E.R., Delgado, J.M.R., Malo, P., and Laviña, M. (in preparation). Enzymatic changes induced by programmed hippocampal stimulation in rabbits.

Gaddum, J.H. (1961). Push-pull cannulae. *J. Physiol.* 155, 1P–2P.

Myers, R.D. (1974). *Handbook of Drug and Chemical Stimulation of the Brain. Behavioral. Pharmacological and Physiological Aspects.* Van Nostrand Reinhold, New York.

Pappenheimer, J.R., Heisey, S.R., Jordon, E.F., and Downer, J. de C. (1962). Perfusion of the cerebral ventricular system in unanesthetized goats. *Amer. J. Physiol.* 203, 763–774.

⁹ Rosenzweig, M.R., Krech, D., Bennett, E.L., and Diamond, M.C. (1968). Modifying brain chemistry and anatomy by enrichment or impoverishment of experience. In: *Early Experience and Behavior* (G. Newton and S. Levine, eds.), pp. 258–297, Charles C. Thomas, Springfield, Ill.

Roth, R.H., Allikmets, L., and Delgado, J.M.R. (1969). Formation and release of NA and DOPA. *Arch. Int. Pharmacodyn.* 181, 273–282.

Sidman, M. (1959). Behavioral pharmacology. *Psychopharmacologia* 1, 1–19.

Skinner, B.F. (1938). *The Behavior of Organisms,* Macmillan, New York.

Skinner, B.F., and Heron, W.T. (1937). Effects of caffeine and benzedrine upon conditioning and extinction. *Psychol. Rec.* 1, 340–346.

Sparber, S.B. (1975). Neurochemical changes associated with schedule-controlled behavior. *Fed. Proc.* 34, 1802–1812.

Stein, L., and Wise, C. (1969). Release of norepinephrine from hypothalamus and amygdala during rewarding medial forebrain bundle stimulation and amphetamine. *J. Comp. Physiol. Psychol.* 67, 189–198.

Winston, J., and Gerlach, J.L. (1971). Stressor-induced release of substances from the rat amygdala detected by the push-pull cannula. *Nature (London) New Biol.* 230, 251–253.

2

New Trends in Behavioral Neurochemistry

F. V. DeFEUDIS

National Center "Ramón y Cajal"
and Autonomous University
Medical School
Madrid, Spain

Regardless of what our philosophy of behavior may be, we must admit that modifications of the external or internal environment are the events that "trigger" behavioral changes. Behavior, which is an expression of learning and memory, is caused by the environment of the organism, though certain limits are set by heredity (Fig. 1). Only through the combined study of the "inside" (biochemistry) and "outside" (behavior) of the organism will the chemistry of behavior be elucidated.

Many studies along this line have already been conducted, and some are discussed in this volume. For example, prolonged exposure of animals to changes in environment can alter their cerebral chemistries as well as their behaviors (e.g., Bennett et al., 1964; Essman, 1971; Valzelli, 1973). Another example involves the use of the "chemitrode" system developed by Delgado (1966) to analyze behavior in the awake animal while sampling simultaneously the chemical and electrical activities of the brain. The collectability from the brain of amino acids, newly synthesized from precursor radioactive glucose, has already been shown with this method (Fig. 2). By using different precursors for labeling

11

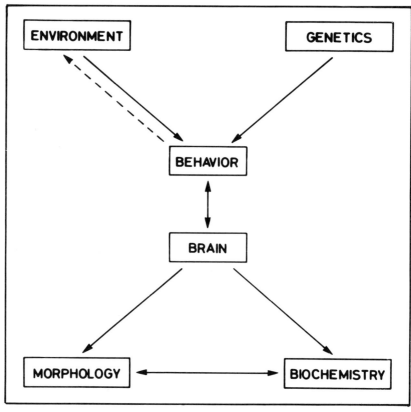

1 Diagram illustrating the interrelationships involved in studies of the neurochemistry of behavior.

2 Amino acid profile and plots of corresponding radioactivity after perfusion of the right caudate nucleus of an anaesthetized rhesus monkey with a solution containing U-^{14}C-D-glucose. An ethanolic extract of 29 mg of caudate nucleus and its corresponding perfusate collected over ½ hr (0.5 ml) were analyzed. Buffer changes and elution times are indicated. The column of resin used for the perfusate was 2 cm shorter than that used for the extract, and hence GABA was eluted earlier, but coincided with the ninhydrin peak for standard GABA. The presence of α-alanine in the perfusate, but not in the extract, indicated that this amino acid was rapidly synthesized and released under the conditions of the experiment. MSO, methionine sulfoxide; ASP, aspartate; GLN, glutamine; ASPN, asparagine; SER, serine; GLU, glutamate; PRO, proline; GLY, glycine; ALA, α-alanine; VAL, valine; ILE, isoleucine; LEU, leucine; TYR, tyrosine; PHE, phenylalanine; GABA, γ-aminobutyric acid (unpublished photograph; data in DeFeudis, Delgado, and Roth, 1970).

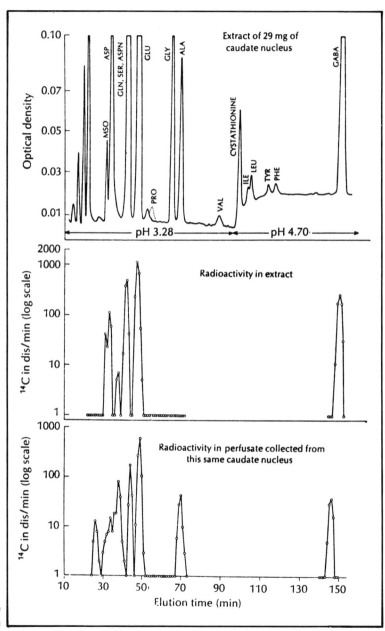

2

and small metabolic compartments of the brain (e.g., glucose
e for the large pool; acetate, glutamate, and leucine for the
.) more information will be gained about the metabolic and
subcellular compartments which contribute to alterations in amino acid
release caused by different experimental conditions. This approach is
based on the pioneering studies performed on metabolic compartmentation in the brain (see e.g., Berl et al., 1961, 1962; Berl and Clarke, 1969).

With respect to genetics and maturation, the nervous systems of
most higher vertebrates are "fully" developed at birth, or shortly thereafter. Therefore, young animals possess many innate neural pathways
which enable them to exhibit reflexes and to emit complex instinctive
patterns of behavior. Studies of these genetically linked behavioral responses and their associated neurochemistries may also provide important
clues about mechanisms of behavior.

Are learning, memory, and behavior concerned with chemical and
anatomic plasticity; i.e., are these accomplished by means of increases
in functional synaptic connections, or by increased numbers of synaptic
boutons, or by increased amounts of neurotransmitter (or associated
enzymes), or by increases in the numbers of receptor sites for transmitters? Perhaps the assumption we must accept in attempting to answer
this question was provided by D. O. Hebb in 1949, i.e., that the neural
change which is induced by experience and which constitutes memory,
is in some way *structural* and *static*. Major questions seem to be: Are
changes in cerebral morphology caused by learning, or by memory, or
by behavior? If these changes in morphology do occur, are they caused
by altered rates of synthesis of preexisting molecules, or by the synthesis
of new species of molecules?

After the works of Ramón y Cajal (1895) and Sherrington (1906),
synapses became a central issue in the study of learning and memory.
It was believed that repeated synaptic use strengthened certain synaptic
connections, and therefore that learning and memory were associated
with neural growth. Later, the idea that other requirements for information storage could be met by macromolecular mechanisms, especially
by proteins, was advanced (see e.g., Bogoch, 1968; Hydén, 1973a).
Macromolecules, such as proteins, or conjugated proteins which are
constituents of neuronal and glial membranes, could undergo conformational changes or increased synthesis during experience, and therefore,
could constitute the mechanism by which electrical impulses are trans-
formed into molecular and cellular growth.

These two approaches regarding the coding or storage of experiential information are not mutually exclusive; one approach emphasizes that

brain growth occurs as a function of experience, whereas the other infers that informational processes occur in macromolecules such as nucleic acids and proteins. Both approaches must apply since the formation of new functional connections, which implies synaptic growth, is necessarily accompanied by molecular changes in cells.

The notion that the brain might exhibit growth through experience was advanced early (e.g., Spurzheim, 1815; Darwin, 1874; Tanzi, 1893; Ramón y Cajal, 1895), and the works of Krech, Bennett, Rosenzweig, and co-workers (see e.g., Bennett and Rosenzweig, 1971) are in accord with this idea. Many other studies have supported the concept that brain growth is related to experience or to function (e.g., Edstrom, 1957; Hubel and Wiesel, 1963; Westrum et al., 1964; Gonatas, 1967; Valverde, 1967, 1968; Geinsmann et al., 1971; Bogolepov and Pushkin, 1975). Therefore, it seems likely that the morphology of the CNS is continuously changing with respect to the level of environmental stimulation received by the organism. The "behavioral code" could be based upon certain specific macromolecular changes that occur in synaptic membranes of large areas of the CNS which are sensitive to environmental modification. Proteins or conjugated proteins appear to be involved.

In support of this contention, it has been shown that cerebral protein synthesis increases with learning. Some proteins whose syntheses were enhanced by learning or training were the acidic species, S-100 and 14-3-2 (e.g., Hydén, 1973a,c), and certain glycoproteins (e.g., Bogoch, 1968). These substances could play roles in neuronal and glial membrane function. Other experiments have indicated that certain inhibitors of protein synthesis (e.g., puromycin) can block some of the effects of training (e.g., Agranoff, 1973).

Evidence that RNA may be involved in learning and memory stems from Hydén's work. Rats trained to obtain food by balancing on a wire perched at an angle exhibited a significant increase in nuclear RNA of neurones and glia of Deiters' nucleus (Hydén and Egyhazi, 1962, 1963). Alterations in RNA base ratios led Hydén to believe that RNA species synthesized during learning differed from those synthesized during control conditions. However, to date, no convincing demonstration exists of a direct relationship between a stimulus and the synthesis of one or more species of RNA. Also, no good evidence exists to support the idea that any brain RNA species goes beyond its role as mediator between DNA and protein synthesis.

Recently, increased emphasis has been placed on the possible roles of glial cells in information processing (e.g., Hydén, 1973b). Both the S-100 proteins (Hydén and McEwen, 1966) and the training-sensitive glycoprotein 10B (Bogoch, 1968, 1972) appear to be glial constituents. Earlier studies showed that the fine structure of cerebral glia could be modified by environmental changes (Diamond et al., 1964, 1966; Altman, 1967), and Krech and co-workers (1966) found an increased number of glial cells in the neocortices of rats raised in "enriched" environments, while the number of neurones remained constant (see review by Bennett and Rosenzweig, 1971). Since glial proteins such as S-100 proteins and glycoprotein 10B are membrane constituents, these substances could be involved in glial–neuronal intercommunication. These results, along with the findings that synapses of many regions of the CNS are closely invested by glial elements, indicate that glia may be involved in synaptic functions which are related to learning, memory, and behavior.

Some of our recent work (F. V. DeFeudis, A. Ojeda, P. Madtes, and P. A. DeFeudis, 1976; in press) is in line with this idea that proteins may be involved in the brain's response to changes in experience which affect learning, memory, and behavior. The effects of long-term changes in environment (which influence behavior and which are expected to influence memory) are being studied in relation to the "binding" of the possible inhibitory transmitters, GABA and glycine, and the protein contents of synaptosomal fractions of mouse brain. An impoverished environment decreased the amount of protein (Fig. 3) and the "binding" of inhibitory transmitter suspects in cerebral synaptosomal structures (Fig. 4). Since no change occurred in the "binding" of amino acids when expressed on a protein basis (Fig. 5), it seems likely that a change in the total amount of synaptic protein or in the total number of synapses had occurred. Less "binding" sites for GABA and glycine and less total protein were present in synaptosomal fractions of the brains of the "aggressive" isolated mice. Gel electrophoretic analyses of peak fractions (Nos. 20 and 25; see Fig. 3) indicated further that this environmentally induced change in synaptic protein occurred uniformly among the 41 protein species that were resolved (F. Conde and F. V. DeFeudis, 1976; to be published). These findings provide an example of an environmentally induced change in cerebral morphology which is reflected in the "binding" of inhibitory transmitter suspects and in the protein contents of synaptic particles and which coincides with dramatic changes in behavior. One might speculate, on the basis of these results that learning, memory, and behavior depend upon the axoplasmic flow of proteins and glycoproteins to synaptic nerve endings (Droz, 1974; Fig. 6), and that a decrease in this supply of macromolecules might account for the decreases in synaptic protein and transmitter "binding" that occurred in the "isolated" mice (Fig. 3-5).

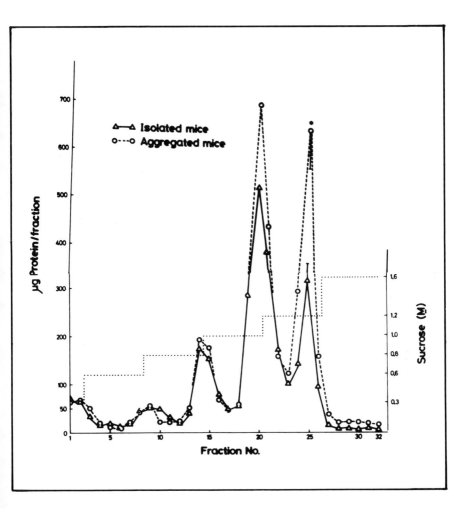

3

Protein contents of "synaptosomal-mitochondrial" fractions of the brains of differentially housed male, Swiss albino mice, as revealed by gradient fractionation. Mice were differentially housed for 6-7 weeks. Points represent mean values for 5 mice in all cases; the level of significance by (Student's t-test) for the difference shown in fraction no. 25 and the relevant standard errors are indicated. Molarities of the gradient are shown on the right-hand ordinate. Fractions are numbered from top to bottom of the tubes (F. V. DeFeudis, et al, 1976).

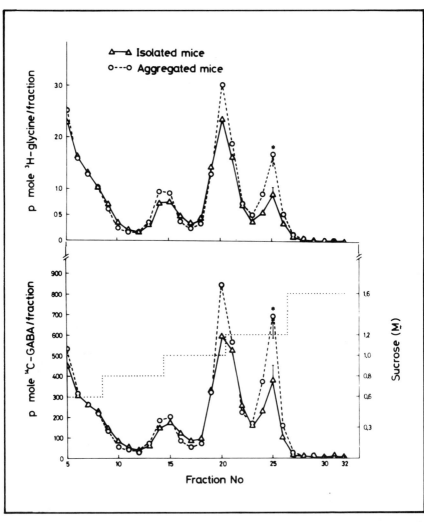

4 "Binding" of ^{14}C-GABA and ^3H-glycine to cerebral "synaptosomal-mitochondrial" fractions of differentially housed male, Swiss albino mice, as revealed by gradient fractionation. Mice were differentially housed for 6-7 weeks. Significant differences (by Student's t-test) and relevant standard errors are indicated; $n = 5$ for each group of mice. Molarities of the gradient are indicated on the right-hand ordinate. Fractions are numbered from top to bottom of the tubes. (F. V. DeFeudis, et al, 1976).

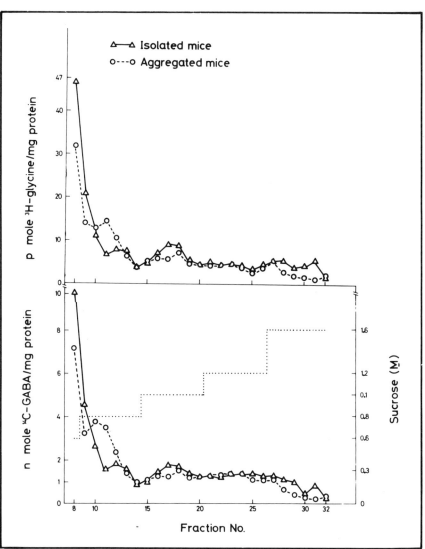

5 Gradient profiles of the "binding" of ^{14}C-GABA and ^{3}H-glycine to cerebral "synaptosomal-mitochondrial" fractions of differentially housed male, Swiss albino mice, expressed in mole/mg protein. Mice were differentially housed for 6-7 weeks. No significant differences existed between fractions from "isolated" and "aggregated" mice. Mean values; $n = 5$ for each point. The sucrose molarities of the discontinuous gradient are indicated. Fractions are numbered from top to bottom of the tubes (F. V. DeFeudis, et al, 1976).

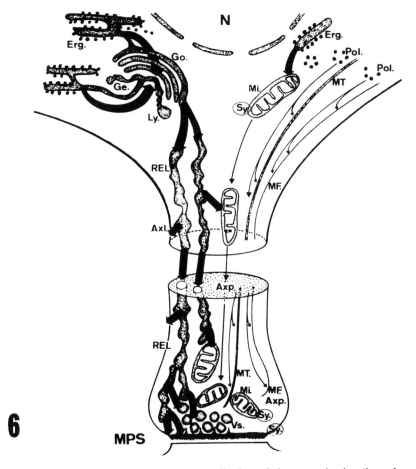

6

Diagrammatic representation of the contribution of the axonal migration of macromolecules to the maintenance of nerve cell processes and synapses. On the left side, the *thick* arrows indicate the part taken by the fast transport. Polypeptide chains synthesized in the ergastoplasm of the Nissl substance (Erg.) are transferred to the Golgi apparatus (Go.) and give rise to protein and glycoprotein sequestered in the smooth endoplasmic reticulum (REL). Passing into the axon with the axonal endoplasmic reticulum (REL), they are transported at a high speed and yield new membrane components to the axolemma (Axl.) and mitochondria (Mi.) They accumulate in the terminal part of the axon and ensure the renewal of constitutents of synaptic vesicles (Vs.) and presynaptic plasma membranes (MPS). Thus the fast transport purveys with new membrane components the logistical support of the conduction and mainly of the transmission of nerve impulses. On the right side, the *thin* arrows point to the elements transported with the slow axonal flow. Polypeptide chains are released from free polysomes (Pol.) into the neuronal cytoplasm and migrate slowly into the axoplasm (Axp.). They may assemble as protein subunits giving rise to microfilaments (MF.) and microtubules (MT.). Mitochondria receive by transfer the great majority of their own proteins; these organelles are displaced along the axon and continue to exhibit a slight synthesis of hydrophobic polypeptides (Sy.). From Droz, 1974.

REFERENCES

Agranoff, B. W. (1973). Biochemical approaches to learning and memory. In: *Macromolecules and Behaviour* (G. B. Ansell and P. B. Bradley, eds.), pp. 143–149, Macmillan, London.

Altman, J. (1967). Postnatal growth and differentiation of the mammalian brain, with implications for a morphological theory of memory. In: *The Neurosciences* (G. C. Quarton, T. Melnechuk, and F. O. Schmitt, eds.), pp. 723–743, Rockefeller Univ. Press, New York.

Bennett. E. L., Diamond, M. C., Krech, D., and Rosenzweig, M. R. (1964). Chemical and anatomical plasticity of brain. *Science* 146, 610–619.

Bennett, E. L., and Rosenzweig, M. R. (1971). Chemical alterations produced in brain by environment and training. In: *Handbook of Neurochemistry* (A. Lajtha, ed.), Vol. 6, pp. 173–201, Plenum Press, New York.

Berl, S., and Clarke, D. D. (1969). Compartmentation of amino acid metabolism. In: *Handbook of Neurochemistry* (A. Lajtha, ed.), Vol. 2, pp. 447–472, Plenum Press, New York.

Berl, S., Lajtha, A., and Waelsch, H. (1961). Amino acid and protein metabolism. VI. Cerebral compartments of glutamic acid metabolism. *J. Neurochem.* 7, 186–197.

Berl, S., Takagaki, G., Clarke, D. D., and Waelsch, H. (1962). Metabolic compartmentation *in vivo*: Ammonia and glutamic acid metabolism in brain and liver. *J. Biol. Chem.* 237, 2562–2569.

Bogoch, S. (1968). *The Biochemistry of Memory; With an Inquiry into the Function of Brain Mucoids.* Oxford Univ. Press, London.

Bogoch, S. (1972). Brain glycoprotein 10B: Further evidence of the "signpost" role of brain glycoproteins in cell recognition, its change in brain tumor, and the presence of a "distance factor." In: *Functional and Structural Proteins of the Nervous System* (A. N. Davison, P. Mandel, and I. G. Morgan, eds.), pp. 39–52, Plenum Press, New York.

Bogolepov, N. N., and Pushkin, A. S. (1975). Submicroscopic changes of cortex nerve cells in chronic mirror epileptic focus in rat. *Brain Res.* 94, 173–184.

Conde, F. P., and DeFeudis, F. V. (1976). Polypeptides of cerebral subcellular fractions of differentially-housed mice. *Submitted for publication.*

Darwin. C. (1874). *The Descent of Man.* 2nd edition, p. 53, Rand McNally, Chicago.

DeFeudis, F. V., Delgado, J. M. R., and Roth, R. H. (1970). Content, synthesis and collectability of amino acids in various structures of the brains of rhesus monkeys. *Brain Res.* 18, 15–23.

DeFeudis, F. V., Ojeda, A., Madtes, P., and DeFeudis, P. A. (1976). "Binding" of GABA and glycine to synaptic fractions of the brains of differentially-housed mice; evidence for morphological changes. *Exptl. Neurol.* In press.

Delgado, J. M. R. (1966). Intracerebral perfusion in awake monkeys. *Arch. Int. Pharmacodyn.* 161, 442–462.

Diamond. M. C., Krech, D., and Rosenzweig, M. R. (1964). The effects of an enriched environment on the histology of the rat cerebral cortex. *J. Comp. Neurol.* 123, 111–119.

Diamond, M. C., Law, F., Rhodes, H., Lindner, B., Rosenzweig, M. R., Krech, D., and Bennett, E. L. (1966). Increases in cortical depth and glia numbers in rats subjected to enriched environment. *J. Comp. Neurol.* 128, 117–125.

Droz, B. (1974). Conclusions générales. *Bull. Acad. Suisse Sci. Méd.* 30, 124–137.

Edstrom, J.-E. (1957). Effects of increased motor activity on the dimensions and staining properties of the neuron soma. *J. Comp. Neurol.* 107, 295–304.

Essman, W. B. (1971). Isolation-induced behavioral modifications: Some neurochemical correlates. In: *Brain Development and Behavior* (M. B. Sterman, D. J. McGinty, and A. M. Adinolfi, eds.), pp. 265–276, Academic Press, New York.

Geinsmann, Y. Y., Larima, V. N., and Mats, U. N. (1971). Changes in neuron dimensions as a possible morphological correlate of their increased functional activity. *Brain Res.* 26, 247–257.

Gonatas, N. K. (1967). Axonic and synaptic lesions in neuropsychiatric disorders. *Nature (London)* 214, 352–355.

Hebb, D. O. (1949). *The Organization of Behavior.* John Wiley & Sons, New York.

Hubel, D. H., and Wiesel, T. N. (1963). Receptive fields of cells in striate cortex of very young, visually inexperienced kittens. *J. Neurophysiol.* 26, 994–1002.

Hydén, H. (1973a). Changes in brain protein during learning. In: *Macromolecules and Behaviour* (G. B. Ansell and P. B. Bradley, eds.), pp. 3–26, Macmillan, London.

Hydén, H. (1973b). Nerve cells and their glia: Relationships and differences. In: *Macromolecules and Behaviour* (G. B. Ansell and P. B. Bradley, eds.), pp. 27–50, Macmillan, London.

Hydén, H. (1973c) RNA changes in brain cells during changes in behaviour and function. In: *Macromolecules and Behaviour* (G. B. Ansell and P. B. Bradley, eds.), pp. 51–75, Macmillan, London.

Hydén, H., and Egyhazi, E. (1962). Nuclear RNA changes during a learning experiment in rats. *Proc. Nat. Acad. Sci. USA* 48, 1366–1373.

Hydén, H., and Egyhazi. E. (1963). Glial RNA changes during learning experiment in rats. *Proc. Nat. Acad. Sci. USA* 49, 618–624.

Hydén, H., and McEwen, B. (1966). A glial protein specific for the nervous system. *Proc. Nat. Acad. Sci. USA* 55, 354–358.

Krech, D., Rosenzweig, M. R., and Bennett, E. L. (1966). Environmental impoverishment, social isolation and changes in brain chemistry and anatomy. *Physiol. Behav.* 1, 99–104.

Ramón y Cajal, S. (1895). *Les Nouvelles Idées sur la Structure du Système Nerveux chez l'Homme et chez les Vertébrés,* edition Francaise revue et augmentée par l'auteur, L. Azoulay, trans., p. 79, Reinwald, Paris.

Sherrington, C. S. (1906). *The Integrative Action of the Nervous System.* Yale Univ. Press, New Haven.

Spurzheim, J. G. (1815). *The Physiognomical System of Drs. Gall and Spurzheim.* 2nd Edition, pp. 554–555, Baldwin, Cradock and Joy, London.

Tanzi, E. (1893). I fatti e la induzione nell'odierne istologia del sistema nervoso. *Riv. Sper. Freniat.* 19, 149.

Valverde, F. (1967). Apical dedritic spines of the visual cortex and light deprivation in the mouse. *Exp. Brain Res.* 3, 337–352.

Valverde, F. (1968) Structural changes in the area striata of the mouse after enucleation. *Exp. Brain Res.* 5, 274–292.

Valzelli, L. (1973). The "isolation syndrome" in mice. *Psychopharmacologia* 31, 305–320.

Westrum, L. E., White, L. E., and Ward, A. A. (1964). Morphology of the experimental epileptic focus, *J. Neurosurg.* 21, 1033–1044.

3

Amine Regulation of Protein Synthesis in Retrograde Amnesia[1]

W. B. ESSMAN
SHIRLEY G. ESSMAN

*Queens College of
the City University of New York
Flushing, New York*

A. INTRODUCTION

The elevation of brain 5-hydroxytryptamine (5-HT) in close tempo-ral proximity with a training experience has been shown to lead to the development of a retrograde amnesia for the behavior based on such training (Essman, 1968, 1970a, 1973). This effect has been produced in rodents by treatment with agents that elevate endogenous 5-HT level, such as electroconvulsive shock (Essman, 1968), as well as by 5-HT given intracranially in quantities sufficient to effect a doubling of the tissue level of this amine (Essman, 1972). As a presumed mediator of a retrograde amnesic effect, increased 5-HT in rodent brain also bears

[1]The work represented in this paper was supported in part by a grant from the Council for Tobacco Research, U.S.A. We wish to acknowledge the technical assistance provided by B. Kornreich, J. Traum, R. Rosenthal, and J. Cusumano.

upon the memory-related process of cerebral protein synthesis (Essman, 1970b). Aside from the memory-disruptive effects observed following inhibition of protein synthesis with centrally administered antibiotics (Agranoff, 1969; Barondes, 1970), 5-HT also reduces the rate at which amino acid precursors are incorporated into cerebral proteins—both generally, and at the subcellular level (Essman, 1970b, 1971).

A number of drugs have been shown to interact with amnesia-producing stimuli such that the behavioral effects of the latter are attenuated. Among these are uric acid which, when administered prior to a training-electroconvulsive shock (ECS) contingency, reduced the incidence of ECS-induced retrograde amnesia in mice. The effect of ECS on brain 5-HT, which appears to be implicated in the production of an amnesic effect, was altered such that ECS when given to uric acid treated mice, did not further elevate forebrain 5-HT content (Essman, 1970c). The purpose of this communication is to report several studies in which the interaction of uric acid and 5-HT were considered with regard to: (1) ECS-induced retrograde amnesia for passive avoidance behavior in the mouse; (2) changes in 5-HT content and disposition at the tissue and cellular level; and (3) effect upon cerebral protein synthesis.

Since uric acid has been shown to reduce the amnesic effect of ECS and to prevent ECS-induced changes in brain 5-HT metabolism, it was anticipated that the amnesic effect of ECS and its inhibition of cerebral protein synthesis could be antagonized by uric acid. The age-specific effects of ECS (Essman, 1970a) and of intracranial 5-HT (Essman, 1972) in the production of retrograde amnesia have been shown to depend on a sustained elevation of cellular levels of 5-HT; failure of this effect to occur, caused either by a failure of 5-HT to achieve cellular distribution or by a rapid 5-HT catabolism, prevented both the retrograde amnesic effect and the alterations in cerebral protein synthesis.

The administration of a single posttraining ECS to mice produces a retrograde amnesia. This effect can be modified by treatment with several agents which may be capable of modifying changes induced by ECS on brain 5-hydroxytryptamine concentration, turnover, and/or uptake. To illustrate the behavioral and neurochemical parallels inherent in the relationship between retrograde amnesia and serotoninergic mechanisms, we have chosen in the experiments herein one agent, uric acid, a methyl xanthine which acts as a mild central nervous system stimulant.

B. RETROGRADE AMNESIA: EFFECTS OF URIC ACID

In an initial experiment, groups of CF-1S strain mice were given a single intraperitoneal injection of either 0.25 ml of 0.9% NaCl or 0.5 – 20 mg/kg doses of uric acid dissolved in an equivalent volume of NaCl. One hour following these treatments, all animals were trained on a single trial passive avoidance task which provided for the acquisition of a stable response. Specifically, the mice were trained by individual placement into a clear Lucite vestibule adjoining a larger darkened chamber, the floor of which consisted of grids wired in series through a cam-operated grid scrambler to a 400 V power supply. Completion of the circuit between any two adjacent grids activated a 3-sec, 2.0 mA footshock. The time interval between entry from the vestibule into the larger chamber through a 5-cm hole was timed and usually occurred within 10 sec after placement of the animal into the vestibule. Ten seconds following chamber entry and footshock, each mouse was given a single transcorneal ECS (20 mA, 200 msec, 400 V), which produced a full clonic-tonic convulsion. All animals were tested 24 hr later for retention of the conditioned avoidance response; this consisted of placing each mouse into the outer vestibule and again timing its entry into the larger chamber in which each had previously received footshock. A conditioned avoidance response was defined by failure to leave the vestibule by 180 sec. Absence of any conditioned avoidance behavior, or retrograde amnesia, was defined by an exit from the vestibule into the larger chamber within a 10-sec interval; approximately the same criterion as had served for the earlier unconditioned responses. By using this criterion, it was possible to distinguish between those mice that showed retention of the passive avoidance response (180 sec) and those that showed amnesia (10 sec) for such behavior.

Figure 1 indicates that 90% of the mice that were pretreated with saline showed evidence of retrograde amnesia. This finding is consistent with earlier reports of retrograde amnesia produced by ECS in rodents (Essman, 1966, 1967, 1968). Animals pretreated with uric acid showed a decreased incidence of retrograde amnesia. This effect was particularly striking at doses of 2 mg/kg and above, representing a 40% or less incidence of ECS-induced retrograde amnesia. This finding suggests that pretreatment with uric acid antagonizes the amnesic effect of ECS.

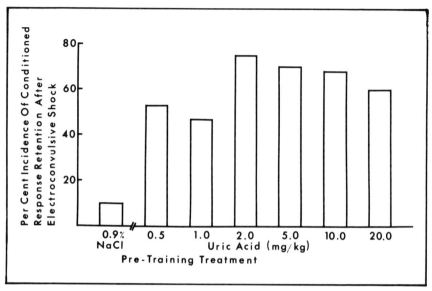

1 Percent incidence of conditioned response retention (testing trial response latency > 180 sec) after electroconvulsive shock in mice pretreated with either saline or several doses of uric acid.

C. ECS-INDUCED SEROTONINERGIC CHANGE: EFFECTS OF URIC ACID

An issue relating to the apparent uric acid mediated antagonism of ECS-induced amnesia concerns the extent to which the same treatment can modify such changes as may occur in brain 5-hydroxytryptamine metabolism. As previously shown (Essman, 1973), a single ECS sufficient to produce a retrograde amnesia, increased brain 5-HT content and the concentration of its metabolite, 5-HIAA, modifying, in addition, its turnover in brain. In the present study, mice pretreated with either 0.9% NaCl or 1 mg/kg of uric acid were given a single ECS of 20 mA 60 min later. Twenty minutes following such treatment, the forebrain tissue was removed and assayed for 5-HT and 5-HIAA concentration (Welch and Welch, 1969). In a parallel study, brain tissue was obtained under identical experimental conditions, following the administration, at several timed intervals, of 20 mg/kg of tranylcypromine. This procedure, which utilized a steady-state kinetics method (Tozer et al., 1966), the accumulation of 5-HT, and the decrement of 5-HIAA following monoamine oxidase inhibition, yielded rate constants from which turn-

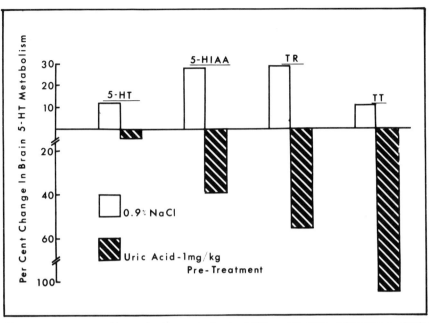

2   Percent change in brain 5-HT metabolism following a single electroconvulsive shock. Mice were pretreated with either saline or uric acid.

over was calculated. The results of this study are summarized in Fig. 2. It should be noted that uric acid reversed all of the changes produced by ECS, thereby leading us to postulate that the antagonism of ECS-induced amnesia by uric acid might represent the reversal of a series of metabolic events classically associated with such disruption of the memory consolidation process.

The foregoing postulate is supported by some prior experimental evidence; for example, it has been shown that age-related determinants of amnesia susceptibility parallel (1) endogenous levels of forebrain 5-HT as a function of age, and (2) the degree to which ECS elevates brain 5-HT as a function of age. Resistance to the amnesic effect of ECS, as indicated by these studies, appears to be associated with low endogenous 5-HT levels and to a relative resistance to elevation by ECS (Essman, 1972).

Perhaps a more significant aspect of the relationship between amnesic agents and brain 5-HT metabolism concerns cerebral protein synthesis. There is ample experimental and theoretical evidence to suggest a relationship between the process of memory consolidation and macromolecular events, implicating most probably a synthesis of brain proteins. It has been shown previously that manipulation of cerebral 5-HT concentration has a rather direct influence upon protein synthesis; namely, (1) low endogenous 5-HT level as a function of age is associated with high levels of protein synthesis (Essman, 1972), (2) depletion of forebrain 5-HT or diurinal variations in forebrain 5-HT (Essman, 1975) are associated with differences at the rate of protein synthesis—higher rates of the latter being associated with lower concentrations of the former, (3) intracranial administration of exogenous 5-HT has been shown to inhibit protein synthesis (Essman, 1972).

D. AGE-RELATED DIFFERENCES IN RETROGRADE AMNESIA: EFFECTS OF URIC ACID

Inasmuch as age-related differences in the susceptibility of the amnesic effects of both ECS (Essman, 1970a) and intrahippocampal 5-HT (Essman, 1972) has been demonstrated and since regional differences in protein synthesis (Essman, 1973) as well as effects of specific amnesic agents were also shown to be age related (Essman, 1972), several experiments were undertaken to investigate the effects of uric acid on (1) the susceptibility to retrograde amnesia as a function of age, and (2) the effects of uric acid upon amnesic stimulus-induced inhibition of protein synthesis as a function of age.

Groups of male CF-1S strain mice were selected at ages ranging from 15 to 30 days. This choice was dictated by previous studies (Essman, 1970a) which showed that by 15 days of age, this strain of mouse was capable of acquiring a relatively stable conditioned avoidance response and was also susceptible to the effects of transcorneal ECS. Mice in each age group were given an intraperitoneal injection of either 0.9% NaCl or an equivalent volume of 1 mg/kg of uric acid. Sixty minutes later, a single training for the establishment of a passive avoidance response was given, and this was followed within 10 sec by the administration of a single transcorneal ECS. Twenty-four hours later, all animals were tested for retention of the conditioned avoidance behavior.

The two major observations made during these experiments have been summarized in Fig. 3. First, saline-treated mice of 20 days of age, or older, showed a predictably high incidence of ECS-induced retrograde amnesia; 15-day-old mice, and to a lesser extent, 16-day-old mice also showed an ECS retrograde amnesia; mice of 17 days of age, however, showed a marked resistance to the amnesic effect of ECS, with

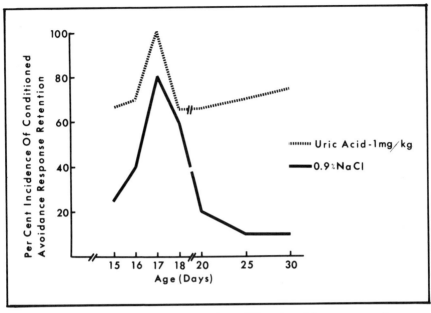

3   Percent incidence of conditioned avoidance response retention (testing trial latency > 180 sec for mice pretreated with saline or uric acid as a function of age).

only a 20% incidence of amnesia resulting for this age group. This finding is again consistent with earlier data that have indicated that 17-day-old CF-1S mice are highly resistant to both ECS- and 5-HT-induced amnesia. Among the uric acid treated mice, it is clear that the maximum degree of retrograde amnesia shown for any age group was 35%, with as little as 15% amnesia being exhibited by 30-day-old mice. It is also apparent that 17-day-old mice, studied under the conditions heretofore cited, showed no retrograde amnesia whatsoever if they were pretreated with uric acid. Likewise, these data indicate that the 17-day-old animals differed significantly ($p<0.01$) from older and younger animals in their ssusceptibility to the amnesic effects of ECS.

E.   AGE DIFFERENCE IN PROTEIN SYNTHESIS: EFFECTS OF URIC ACID

In previous experiments we have shown that there are a number of features which characterize the 17-day-old CF-1S mouse. These include: (1) low endogenous levels of 5-HT; (2) increased rate of protein synthesis in the cerebral cortex, basal ganglia-plus-diencephalon and midbrain; (3) reduced accumulation of exogenous 5-H; and (4) reduced susceptibility to the retrograde amnesic effects of ECS or intra-hippocampal 5-HT injection (Essman, 1972, 1973, 1974). In an attempt to relate some of these findings to the present data, the effects of uric acid upon ECS-induced inhibition of protein synthesis was investigated in male, 15 to 30-day-old CF-1S mice.

Mice of various age groups were given a single intraperitoneal injection of 0.9% NaCl or 1 mg/kg of uric acid. Sixty minutes later, a single transcorneal ECS was administered coincident with the intra-cranial injection of ^{14}C-leucine. The injection was given 1 mm lateral to the midline and 5 mm posterior to the intraocular line and at a depth of 3 mm. Ten microliters of 0.001 μM of ^{14}C-leucine (60 μCi/mole) was injected. It should be noted that this concentration is well below the endogenous level of this amino acid in whole brain and therefore would not be expected to alter the specific activity of whole brain proteins. At this concentration, the effects of isotopic dilution might also be precluded. After a 15-min labeling pulse, the animals were killed and the brain tissue was removed and frozen in Freon. Their cerebral cortices were homogenized in a solution of 0.25 M sucrose, 0.05 M Tris-HCl buffer, pH 7.5, 0.0025 M KCl and 0.001 M MgSO$_4$, in which .075 g/l of whole leucine was dissolved. Determinations of radioactivity were carried out in the material precipitated from aliquots of homogenates by addition of 5 ml of 5% tricholoroacetic acid (TCA). Precipitates were washed twice with cold 5% TCA and then hydrolyzed in 5% TCA at 90°C for 20 min. After washing the precipitate once more, it was extracted with an ethanol–ether mixture (2:1, v/v), and then dissolved in concentrated formic acid. Radioactivity was determined after addition of 15 ml of scintillation fluid.

Results are expressed as cpm ^{14}C-leucine/mg protein (Fig. 4). It is apparent that a 35 to 48% inhibition of protein synthesis occurred in mice of all ages, except for those in which this procedure was carried out at 17 days of age. The statistically significant reduction of ^{14}C-leucine incorporation into forebrain proteins observed in the 17-day-old animals (10%) contrasted markedly with the greater degree to which ECS inhibited protein synthesis in mice at other ages. Pretreatment of animals with uric acid significantly reduced ($p<0.01$) the magnitude of ECS-induced inhibition of protein synthesis. The maximum extent to which incorporation of ^{14}C-leucine into protein occurred was still less than 25%. A significant difference in ^{14}C-leucine incorporation between saline-

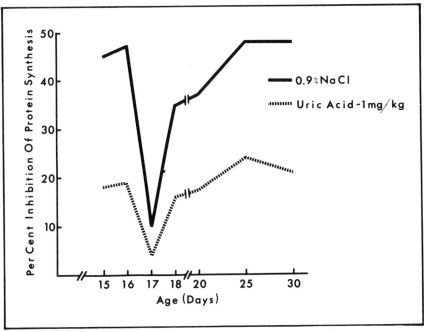

4   Percent incidence of cerebral protein synthesis produced by a single electroconvulsive shock in mice of different ages pretreated with saline or uric acid.

treated and uric acid-treated mice was observed as a function of age ($p<0.02$). Although not statistically significant, it is apparent that of all animals treated with uric acid, only 17-day-old animals showed a reduction in the magnitude of ECS-induced inhibition (3% in contrast to 10% for saline-treated mice at this age). Therefore, from the data shown it is patently clear that in addition to reducing the amnesic effect of ECS, pretreatment with uric acid antagonizes the ECS-induced inhibition of cerebral protein synthesis. One might ask at this time the particular reason for choosing the cerebral cortex as a site to illustrate this effect. Previous studies have shown that introduction of the labeled amino acid into subcortical regions still provides for appreciable amino acid activation and appreciable rates of incorporation into proteins of the cerebral cortex. Furthermore, inhibition of protein synthesis has been demonstrated in the cerebral cortex as a consequence of ECS. The subcellular sites at which these effects occur have also been identified in the cerebral cortex (Essman, 1974).

F. SYNAPTOSOMAL PROTEIN SYNTHESIS: EFFECTS OF ECS AND URIC ACID

Several *in vitro* experiments were performed to further assess the effects of ECS upon regional protein synthesis at a subcellular level. Synaptosomal preparations, derived by several methods, from several regions of the mouse brain were employed.

Male CF-1S were given a single transcorneal ECS (20 mA) and killed 20 min later by cervical dislocation. Control mice received sham-ECS and identical subsequent treatment. The brain tissue was rapidly removed, dissected into several regions and maintained throughout experiments at 2°C. Cerebral cortex, basal ganglia-plus-diencephalon, midbrain, and cerebellar cortex were studied. The pooled tissue for each brain region was homogenized in cold 0.32 M sucrose to provide a final concentration of 10% (w/v).

Synaptosomes were prepared utilizing variations of two basic procedures (Whittaker, 1969): tissue from the cerebral cortex and basal ganglia-plus-diencephalon was homogenized; the homogenate was centrifuged at 10^4 g min; and the resulting supernatant fraction was centrifuged at 5×10^5 g min yielding a pellet containing myelin, synaptosomes, and mitochondria. This pellet, suspended in a 0.32 M sucrose solution, was layered onto a density gradient consisting of equal volumes of 0.8 M sucrose and 1.2 M sucrose and was centrifuged at 6×6^6 g min. The material between 0.8 M and 1.2 M was removed, resuspended, and layered onto a sucrose density gradient of 1.14 M and 1.18 M, respectively, for the cerebral cortex and for the basal ganglia-plus-diencephalon. After centrifuging for 6×10^6 g min, the synaptosomal fraction extracted from the cerebral cortex was removed from the layer just below 1.14 M; material between 1.14 M and 1.18 M represented the synaptosomal fraction of the basal ganglia-plus-diencephalon. Separation of synaptosomes from the midbrain structures and cerebellar cortex consisted of transfering the pellet resulting from centrifugation (10^4 g min) of the 10% homogenate to a 24% Ficoll solution and recentrifuging at 3×10^5 g min. The resulting supernatant was combined with an equal volume of 0.32 M sucrose and centrifuged at 8×10^4 g min. The resultant pellets were resuspended in 0.32 M sucrose and layered onto density gradients consisting of equal volumes of 1.0 M sucrose and 1.4 M sucrose. These gradients were centrifuged for 6×10^6 g min. The band just below 1.0 M sucrose constituted the synaptosome fraction of the pooled midbrain material. For the cerebellar synaptosomes (mossy fiber terminals) the layer just above 1.4 M sucrose was taken.

The specific synaptosomal elements from each brain region, as prepared by the above procedures, were taken to a volume of 0.5 ml and added to an incubation medium consisting of 50 mM Tris-KCl buffer, pH 7.5 100 mM KCl, 5 mM MgSO$_4$, 2mM ATP, 0.1 mM GTP, 15 mM creatine phosphate, 50 μg/ml creatine phophokinase (30 units/mg). To this mixture were added 0.1 μmole/ml of ^{14}C-leucine (0.1 ml) and 0.3 ml of microsomal protein (fraction obtained during the synaptosome preparation) after which incubation was carried out at 37°C with constant agitation for 15 min. Each set of synaptosomal fractions was incubated as described above in the presence of 10^{-5} M uric acid and under control conditions (addition of 0.9% NaCl). The reaction was stopped by the addition of 5 ml of 5% TCA and the resulting medium was centrifuged and washed twice with 5 ml of 5% TCA. The washed precipitate was taken up in concentrated formic acid combined with scintillation fluid for the determination of the incorporated radioactivity.

The protein contents of all synaptosomal fractions were determined, and the final specific activities (cpm/mg protein) were assessed.

The data from the present experiments have been graphed by comparing the difference between ECS and non-ECS treated mice with respect to the rate of ^{14}C incorporation into respective synaptosomal fractions. The synaptosomal derivatives from each region were treated *in vitro* with either uric acid or saline. In accord with earlier observations, a significant degree of inhibition of synaptosomal protein synthesis was observed for three brain regions (cerebral cortex; basal ganglia-plus-diencephalon; midbrain). The use of tissue from the cerebellar cortex of ECS-treated mice in *in vitro* studies of synaptosomal protein synthesis did not yield remarkable reductions. In the presence of physiological quantities of uric acid, the magnitude of inhibition of protein synthesis in synaptosomes derived from ECS-treated mice was significantly reduced ($p < 0.01$) for all regions except cerebellar cortex. Greater than 50% reduction of the magnitude of inhibition was achieved by addition of uric acid during the *in vitro* incubation. These findings support earlier *in vivo* data which indicated that ECS inhibited cerebral protein synthesis as well as the protein synthesis of synaptosomal fractions. The present data indicate further that such inhibition of protein synthesis, as effected *in vivo,* can be significantly attentuated *in vitro* by the presence of small amounts of uric acid.

G. REGIONAL SYNAPTOSOME MORPHOLOGY

Utilizing various regions of the brain, synaptosome fractions were obtained from which small aliquots were removed and mixed with cold

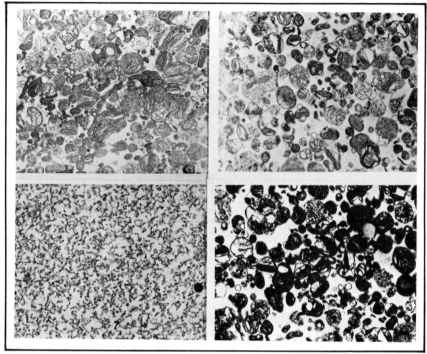

5 Representative sections of synaptosome fractions prepared from different regions of the mouse brain. Transmission electron micrographs represent glutaraldehyde-fixed, uranyl lead acetate prepared sections as follows: upper left, cerebral cortex; upper right, basal ganglia and diencephalon; lower left, midbrain; lower right, cerebellar cortex.

2.5% glutaraldehyde solution (pH 7.4) in cellulose nitrate tubes. The samples were centrifuged at 50,000 rpm for 30 min. The resulting pellets were washed in 0.1 M phosphate buffer (pH 7.4) and then with 95% ethanol (v/v), a small amount of which was left just above the pellets. These "fixed" fractions were then prepared for transmission or scanning electron microscopy for verification of fraction purity. Representative sections from transmission electron micrographs of these synaptosomal fractions are shown in Fig. 5. A representative scanning electron micrograph of an individual synaptosome derived from mouse cerebral cortex is also shown (Fig. 6). It is apparent that relatively pure fractions were obtained which are comparable to the purity of other fractions previously reported. It seems appropriate to mention that differences in synaptosome density and sedimentation properties exist, depending on the region of the mouse brain selected for study. In this regard, it would certainly not be appropriate to apply specific densities for cerebral cortex to subcortical brain regions; nor would it be appropriate to utilize a single existing density gradient methodology and assume uniformity or purity of resulting fractions if this method were utilized on whole brain. We have chosen to rely upon morphological markers for synaptosomal purity inasmuch as their validity for such experiments has been confirmed by a high degree of consistency from experiment to experiment.

H. SYNAPTOSOMAL 5-HYDROXYTRYPTAMINE UPTAKE: EFFECTS OF ECS AND URIC ACID

Apart from the increase in brain 5-HT concentration and the reduction in the rate of brain protein synthesis produced by ECS, this treatment also alters 5-HT uptake into synaptosomes. This effect has been observed in several brain regions (Essman, 1974). Significant reductions

6 Representative scanning electron micrograph of an individual synaptosome derived from mouse cerebral cortex. This view shows a glutaraldehyde-fixed sample, shadowed with gold, at a magnification of 6000×.

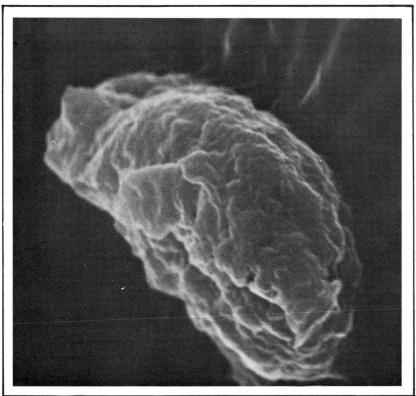

of 5-HT uptake by synaptosomes derived from cerebral cortex, midbrain, and cerebellum have been shown after administration of ECS.

In an effort to explore further the relationship between ECS and synaptosomal 5-HT uptake in the present context, male CF-1S mice were given a single transcorneal ECS or sham ECS. Animals were killed 15 min following such treatment and tissue was removed from the cerebral cortex, midbrain structures, and cerebellar cortex. Synaptosomes were prepared as previously described from pooled tissue of these selected regions; utilizing 0.1 ml of the representative synaptosomal fractions, 0.1 ml of a 0.04 μCi/ml solution of ^{14}C-5-HT, and 3.7 ml of Krebs–Henseleit solution, containing 1.25×10^{-2} M pargyline; incubation was conducted for 15 min at $37°C$. The reaction mixture was brought to a final volume of 4 ml by addition of 0.1 ml of 0.9% NaCl or 10^{-5} M uric acid in NaCl. The incubation was maintained by a 95% O_2/5% CO_2 gaseous phase on a metabolic shaker, after which the reaction solution was centrifuged at 35,000 rpm for 20 min to stop the reaction. Resulting precipitates were washed twice with 3 ml of 0.9% NaCl and the washed precipitates were then taken up in 2 ml of absolute (100%) ethanol. Protein determinations were carried out on representative aliquots. The amounts of radioactivity incorporated into synaptosomal fractions were determined by addition of scintillation fluid and subsequent counting in a scintillation spectrometer. The results were expressed as cpm of ^{14}C-5-HT/mg protein.

Differences in synaptosomal 5-HT uptake produced by ECS have been summarized in Fig. 7. It is clear that significant reductions in the *in vitro* uptake of 5-HT by synaptosomes is related to antecedent *in vivo* ECS treatment. It is also apparent that for at least two regions of the brain from which synaptosomes were derived (cerebral cortex and midbrain) uric acid blocked the ECS-mediated reduction in synaptosomal 5-HT uptake. This effect was not evident with synaptosomes derived from the cerebellar cortex. These findings·again suggest that the *in vitro* effects of uric acid tend to antagonize or to block ECS-induced metabolic changes. In the case of 5-HT uptake by the synaptosome, it may be theorized that the effects of ECS modify either the storage properties of this amine, thereby increasing the concentration gradient against which transport into the synaptosome occurs, or alter the permeability of the external synaptosomal membrane to the influx of 5-HT molecules.

I.   SEPARATION OF NEURONAL PERIKARYA AND GLIA

Several questions raised in the previous experiments constitute the groundwork for subsequent studies conducted in this laboratory, in

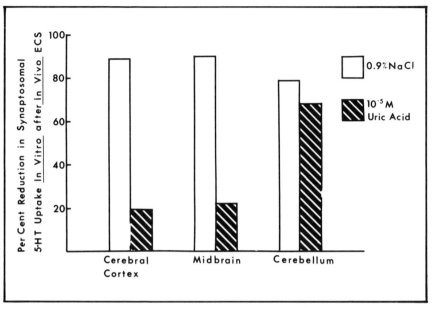

7 ECS-induced reduction in synaptosomal 5-HT uptake in the cerebral cortex, midbrain and cerebellar cortex. *In vitro* uptake was determined in the presence of either saline or uric acid.

which cell-specific differences in metabolic events associated with amnesic stimuli were considered. One such question, previously considered, disclosed the observation that 5-HT uptake is inhibited *in vitro* as a consequence of ECS administration *in vivo*. It was of interest, therefore, to determine whether or not cellular differences over several brain regions might also reflect the *in vitro* differences resulting from *in vivo* treatment. In order to extend these studies, it was desirable to utilize neuronal perikarya and their adjacent glia. Consequently, a method for the separation of neuronal cell bodies from their adjacent glia was developed, taking into account differences in particle density, sedimentation properties, and variations in glial cell type and distribution.

The procedure employed in the preparation of tissue containing enriched pools of neuronal cell bodies and adjacent glia was delineated as follows: pooled tissue extracted from each brain region was chopped to a uniform mince and put into a 10 mM KH$_2$PO$_4$–NaOH buffer (pH, 6.0) to which a solution of 5% glucose, 5% fructose, and 1% BSA (Fraction V) was added. The minced mixture was forced through a 153 mesh nylon filter and then filtered through a 250 mesh stainless steel screen. The resultant filtrate was brought to 3 times its volume by the addition

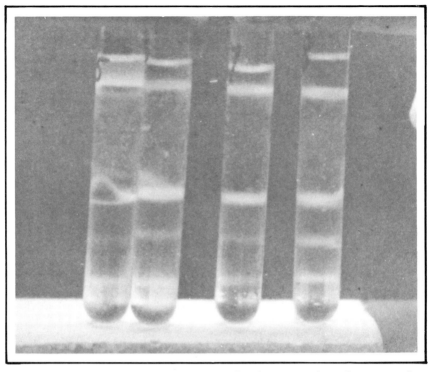

8 Density gradient prepared for the preparation of a neuronal cell body-enriched fraction and a crude glial fraction (see text).

of glucose–fructrose–BSA (GFA) solution and centrifuged at 63×10^2 g min. The pellet produced by this procedure was resuspended in two volumes of the GFA buffer medium and layered over a gradient consisting of 2.0 M sucrose, 1.55 M sucrose, 1.35 M sucrose, and 0.9 M sucrose, all of which were prepared in GFA medium. After centrifugation at 33×10^3 g min, the gradient contained four basic fractions corresponding to bands in the sucrose media (see Fig. 8). The layer just above 2.0 M sucrose was removed for preparation of a fraction enriched in neuronal cell bodies, and the layer just above 1.3 M sucrose was removed for further purification of the glial fraction. The basic neuronal fraction was centrifuged at 33×10^3 g min after being applied to a sucrose gradient consisting of equal volumes of 2.0 M sucrose and 1.2 M sucrose (for cerebral cortex and basal ganglia-plus-diencephalon, respectively), 1.16 M sucrose (midbrain), or 1.64 M sucrose (cerebellar cortex). The material which formed a band just above the 2.0 M sucrose layer in each fraction represented the purified fraction enriched in neuronal cell bodies. The crude glial fraction, also obtained by this procedure, was brought to a volume of 7 ml by addition of a GFA buffer medium and layered onto a gradient consisting of equal volumes of 1.4 M sucrose and 0.9 M sucrose, the sucrose being brought to appropriate molar concentration in the GFA medium. This gradient, centrifuged for 66 $\times$ 10³ g min, provided 3 bands.

The final density gradient with banded particles is shown in Fig. 9. The two later bands, one just above 0.9 M sucrose and one below this layer, corresponding approximately to a density of 1.15 M, represented the two major populations of glia (oligodendrocytes, astrocytes) isolated by this procedure.

Aliquots from each of the neuronal cell body- or glia-enriched fractions prepared from the brain regions of interest were fixed as previously described and prepared for scanning electron microscopy (SEM). The results of SEM analysis of these fractions are summarized (Figs. 10, 11, and 12). These represent, respectively, a characteristic single neuronal cell body, a single oligodendrocyte, and a single astrocyte.

9 Density gradient prepared for isolation of a glia-enriched fraction, consisting of oligodendrocytes (above 0.9 M sucrose) and astrocytes (corresponding approximately to 1.15 M sucrose) (see text).

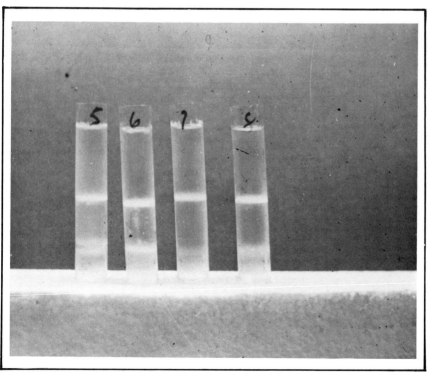

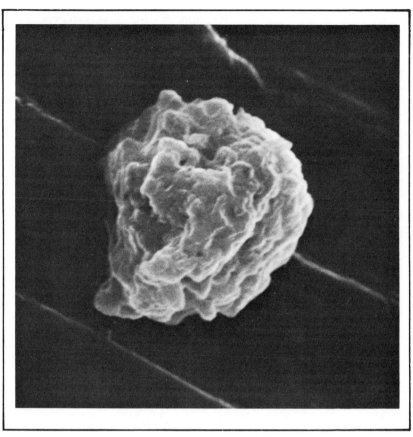

10 Scanning electron micrograph of a single neuronal cell body prepared by fractionation of tissue from the cerebral cortex. Magnification is at 4500×.

J. CELLULAR 5-HT UPTAKE: ECS AND URIC ACID EFFECTS

These experiments were carried out to assess the effects of uric acid upon the *in vitro* uptake of 5-HT by neuronal cell bodies and glia derived from tissue obtained from ECS-treated mice.

Male, CF-1S mice were given either a single transcorneal ECS (20 mA) or sham-ECS. Twenty minutes later, the animals were killed and their cerebral cortices, basal ganglia, diencephala, and cerebellar cortices were dissected out. The tissue for each of these regions was pooled and subjected to the fractionation procedure described above. Utilizing the neuronal cell body and glial fractions obtained, a 0.1 ml aliquot from each fraction was added to 3.7 ml of a medium consisting of: (1) sucrose equimolar to the solution in which the cell fraction was suspended, (2) 0.5 M Tris-HCl buffer, pH 7.5; (3) 0.025 M KCl, and (4) 0.002 M MgSO$_4$, 0.1 ml of ^{14}C-5-HT (1.66 μCi/ml) along with 0.9% NaCl or 10^{-5} uric acid in saline to a final volume of the incubation mixture of 4 ml. The reaction mixture was incubated on a shaker bath for 45 min at 37°C in the presence of a 95% O$_2$/5% CO$_2$ gas mixture. After stopping the reaction by addition of an equal volume of water (0° to 2°C), the mixture was centrifuged at 10,000 rpm for 10 min. The supernatant was mixed with an equal volume of 15% TCA and centrifuged at 12,000 rpm for 10 min. Then, the resulting precipitate was washed twice with 7% TCA, the final precipitate being taken up in 1 ml of 0.32 M sucrose and subsequently counted in scintillation fluid. Protein determinations were carried out on the relevant fractions and the amount of 5-HT taken up by the respective fractions was expressed in terms of the protein concentration.

11 Scanning electron micrograph of single isolated oligodendrocyte derived from fractionation of the mouse cerebral cortex. Magnification is at 4500×.

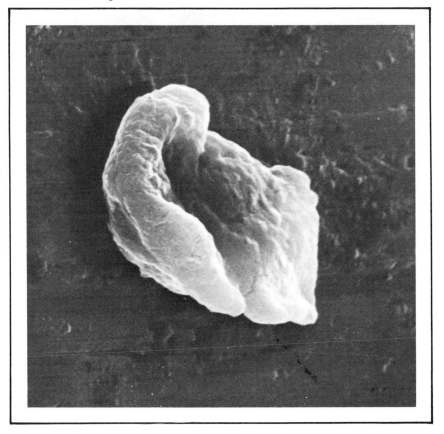

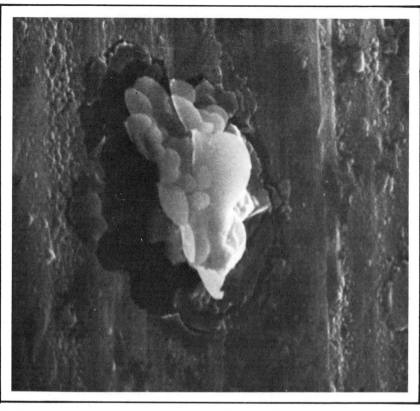

12 Scanning electron micrograph of single isolated astrocyte prepared from mouse cerebral cortex. Magnification is at 4500×.

The results of these studies have been summarized separately for neuronal cell bodies and for glia (Figs. 13 and 14). Both fractions showed some decrease in 5-HT uptake as a consequence of *in vivo* ECS. This was significant in the case of neuronal cell bodies from the basal ganglia and cerebellum and for glia from the basal ganglia-plus-diencephalon. It is also clear that the *in vitro* addition of uric acid to ECS-treated tissue attenuated or reversed the ECS-related reduction in cellular 5-HT uptake. A significant reversal of this phenomonen was seen in neuronal cell bodies of the cerebral cortex, diencephalon, and cerebellum, in which uric acid appeared to markedly facilitate 5-HT uptake. Also, whereas 5-HT uptake was significantly reduced in glia of the basal ganglia-plus-diencephalon, no significant increase in 5-HT uptake was produced following uric acid addition. Therefore, although *in vivo* ECS appears to be related to *in vitro* 5-HT uptake decrements, this effect can be blocked or reversed by the addition of uric acid *in vitro*.

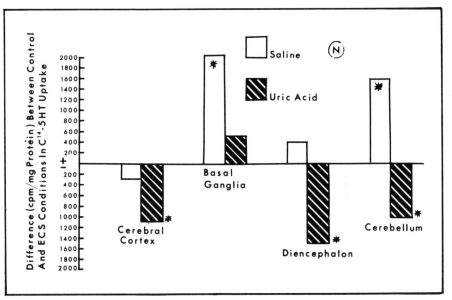

13   Effects of ECS ¹⁴C-5-HT uptake by neurons from several regions of the mouse brain; *in vitro* effects of saline and uric acid.

The precise mechanism by which this interaction occurs still remains a matter of conjecture. It may be speculated, however, that the mechanism resides in a number of potential processes related to the regulation of cellular 5-HT uptake. Perhaps one such process is the possible effects of some unaccounted-for variables of the present experiments upon cyclic nucleotides and their potential role in uptake mechanisms. It is apparent, however, that at least some effect upon cyclic nucleotide availability may be mediated by uric acid since uric acid is a methyl-xanthine which inhibits phosphodiesterase. It is also not absolutely clear why neuronal cell bodies and their adjacent glia were affected similarly by ECS-induced changes in 5-HT uptake and why uric acid reversed these changes. Moreover, inasmuch as it has previously been shown that synaptosomes exhibit the same phenomenon of uptake alteration and reversal as has been demonstrated in neuronal cell bodies and glia, there may well be some lack of specificity for the phenomenon reported or, indeed, said phenomenon may be explained as a function of exogenous 5-HT limited by membrane barriers, the properties of which themselves can be altered by ECS. Allowing the latter as a possibility, it would seem to follow that uric acid is capable of modifying such membrane alterations, as has been described.

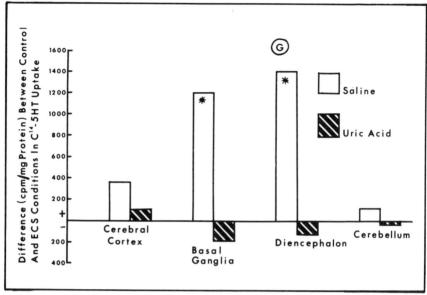

14 Effects of ECS ^{14}C-5-HT uptake by glia from several regions of the mouse brain: *in vitro* effects of saline and uric acid.

K. CELLULAR PROTEIN SYNTHESIS:
ECS AND URIC ACID EFFECTS

We observed that the inhibition of protein synthesis by ECS could be limited by factors which either prevented tissue 5-HT concentration increments or prevented the uptake of 5-HT at sites where proteins were synthesized. One such factor appears to be uric acid which, as we have seen, blocks all effects of induced protein synthesis inhibition at the synapse. It appeared to us that the information accruing from these studies could be augmented by yet another series of studies designed to assess the effects of uric acid on protein synthesis in isolated neuronal cell bodies and glia in several regions of the mouse brain.

Populations enriched with either neuronal cell bodies or adjacent glia were prepared by the previously described method. These populations were obtained from several regions of the mouse brain. The mice employed in this experiment had previously been given either a single transcorneal ECS or sham ECS, the brain tissue being obtained 20 min later. The cell populations were incubated as described above, i.e., utilizing an *in vitro* system for the incorporation of ^{14}C-leucine into cell

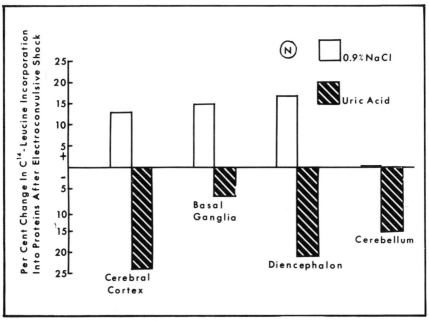

15 Alterations in ^{14}C-leucine incorporation into proteins after electroconvulsive shock; effects of saline or uric acid on *in vitro* protein synthesis by neurons.

proteins. Incubations for cell-enriched populations for each of the specific brain regions (cerebral cortex, basal ganglia, diencephalon, or cerebellum) were carried out in the presence of uric acid (10^{-5} M) or an equivalent volume of 0.9% NaCl for 45 min. On the basis of the counts per minute per milligram protein, the change in leucine incorporation into protein as a consequence of ECS was calculated.

The results pertaining to neuronal cell body-enriched fractions have been summarized in Fig. 15. It should be noted that protein synthesis, as inferred from the ^{14}C-leucine incorporation into proteins, was decreased in the neuronal cell bodies from at least three of the four regions of the brain sampled (cerebral cortex, basal ganglia, and diencephalon); no change was observed for cerebellum. *In vitro* treatment with uric acid reversed the pattern of amino acid incorporation into proteins. Tissue from ECS-treated mice from which neuronal cell bodies were obtained and incubated with uric acid showed increased protein synthesis. This effect was greatest for the cerebral cortex and diencephalon and was most apparent in the same direction in the cerebellar cortex.

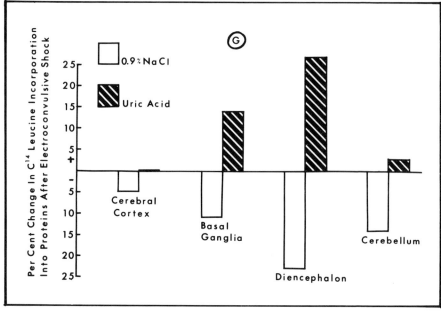

16 Alterations in ^{14}C-leucine incorporation into protein after electroconvulsive shock; effects of saline or uric acid on *in vitro* synthesis by glia.

Results obtained from glia (Fig. 16) contrast quite interestingly with those observed for neuronal cell bodies. Specifically, as a consequence of ECS treatment alone, glia from all regions of brain tissue sampled showed increased rates of amino acid incorporation into proteins. This finding would suggest that ECS, apart from other factors, stimulated protein synthesis in glia. It is of further interest that the greatest increase of leucine incorporation into proteins occurred in glia derived from the diecephalon. It is noteworthy that this is also the region from which neuronal cell bodies showed the greatest ECS-induced decrease in protein synthesis. And, as in the neuronal cell bodies, the *in vitro* effects of uric acid blocked or reversed the *in vivo* effects initiated by ECS. Significantly, protein synthesis in the presence of uric acid was decreased in glia derived from the basal ganglia and cerebellum.

These findings, taken collectively, suggest a most interesting functional reciprocity between neuronal cell bodies and their adjacent glia in several regions of the mouse brain. This reciprocity, which appears to obtain particularly for protein synthesis, bears a striking resemblance to *in vivo* results obtained with other macromolecules similarly affected by ECS. Explicitly, previous studies have shown that nucleic acids decrease in a neuronal population following ECS (Essman, 1966; Essman and Essman, 1969). In contrast to the unidirectional changes in neuronal and glial amine uptake, the effects of ECS upon protein synthesis appear to be bidirectional (decrease in neuronal perikarya; increase in glia). Similarly, the effects of uric acid are also reciprocal and bidirectional. In either case, this methylxanthine blocked or reversed the *in vitro* changes in protein synthesis initiated by ECS.

All of the foregoing studies suggest interrelationships between the amnesic effect of at least one event (ECS), its sequelae—alterations in the concentration and metabolism of 5-HT, 5-HT uptake, and protein synthesis—and the attenuation and/or reversal of specific behavioral and biochemical effects by uric acid. Inasmuch as all of the studies described in this paper have relied on ECS as a medium through which behavioral and biochemical changes were explored and altered by uric acid treatment, *in vivo* as well as *in vitro,* the issue concerning the extent to which the generality of these relationships obtain is raised. Namely, it was useful to inquire experimentally into possibly amnesic agents and to determine for such whether a similar behavior–biochemical–pharmacological consistency prevailed.

L. RETROGRADE AMNESIC EFFECTS OF BUTYRIC ACID

An initial study was concerned with the basic amnesic properties of butyric acid. These experiments were based largely on the rationale that butyric acid, a short-chain fatty acid, has been implicated in the etiology of hepatic coma (Zieve, 1966; Walker and Schenker, 1970); it is capable of binding lecithin in the neuronal membrane and may also complex with several neurotransmitters including 5-HT and norepinephrine (Rizzoli and Galzinga, 1970). One potential result of binding at synaptic sites could be the release of 5-HT. An elevated brain level of 5-HT has been observed in rats during hepatic coma (Baldessarini and Fischer, 1973), and there can be little doubt that one of its concomitants is an increase in circulating butyric acid (Muto and Takahashi, 1965) which has been independently shown to bring about hepatic coma (Samson et al., 1956; Muto et al., 1964) and the EEG pattern characteristic of hepatic coma (White and Samson, 1956).

Serotonin has been further implicated in the encephalopathy of hepatic coma in studies wherein hepatic devascularization in pigs led to increased tryptophan concentration in the hypothalamus, thalamus, caudate, and cortex, increased 5-HT content in the former two regions, and increased 5-HIAA levels in all these regions (Curzon et al., 1973).

The CSF of these animals showed almost a 70% increase in 5-HIAA level with hepatic coma. The level of 5-HIAA in CSF has been shown to increase in hepatic coma (Lal et al., 1975), suggesting an increased brain 5-HT turnover with such encephalopathy (Knell et al., 1974). Further evidence from experimental studies with dogs (Ogihara et al., 1966) has pointed to tryptophan as a causative factor in hepatic encephalopathy, in which there are elevated brain 5-hydroxyindole compounds and an associated decrease in monoamine oxidase activity. We suggest that one aspect of hepatic encephalopathy—the manifestation of episodic myoclonus—is linked to altered brain 5-HT metabolism. This relationship, whether independent of the hepatic encephalopathy induced by butyric acid or part of the syndrome, fits the criteria by which retrograde amnesia may be anticipated (increased 5-HT and 5-HIAA contents and increased 5-HT turnover). It further serves as a basis for predicting biochemical changes consistent with such amnesia, i.e., inhibition of cerebral protein synthesis and reversal thereof by uric acid.

In the present experiments, based upon earlier dose response observations (Cusumano, 1975), an intraperitoneal injection of 22.5 μmoles/g of butyric acid titrated to a pH of 7.0 was administered. At this dose, there was a rapid onset (8.3 $\pm$ 2.7 min) of righting reflex loss, a brief myoclonic episode, and complete coma, lasting for 68.5 $\pm$ 3.5 min. These events appeared in approximately 80% of the animals treated with this regimen. A mortality rate of 10% accounted for the remaining animals. Four hours following butyric acid injection, all animals had completly recovered full sensory and motor capacity and there was no evidence of any residual neurological or toxic effect that would differentiate these animals from saline-treated controls.

In the behavioral portion of this study, groups of animals were given a single training trial in the one-trial passive avoidance response acquisition paradigm previously described. Animals received a conditioning footshock of 2.0 mA for 3 sec upon entry into the larger chamber from the vestibule. A single intraperitoneal injection of either butyric acid or an equivalent volume of 0.9% NaCl was given. These were administered either immediately (0) or at 4, 16, or 64 min following the training trial. Any animal that failed to enter the inner chamber in 12 sec during the training trial was removed from the experiment. Likewise, any animal that failed to lose its righting reflex, to exhibit myoclonus, and to become comatose in 12 min following butyric acid injection was eliminated from the experiment. Twenty-four hours following training, a single testing trial was given to all animals; no shock was delivered on this occasion. Ninety seconds after placement in the outer vestibule, failure of any animal to enter the chamber in which

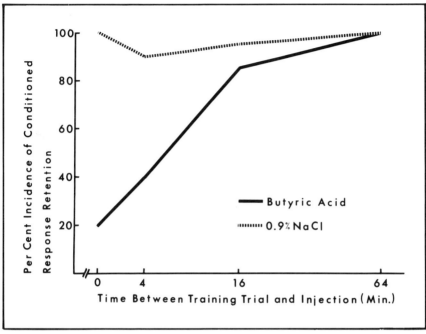

17 Incidence of conditioned response retention (testing trial response latency > 90 sec) as a function oft he time interval between trainin gand injection of saline or butyric acid.

shock had previously been delivered terminated the training trial, and such animals were removed. Any animal that entered the shock chamber within 12 sec after placement in the vestibule was considered to have met the criterion for retrograde amnesia.

The results of this experiment (Fig. 17) demonstrated that the incidence of conditioned response retention is intimately related to time factors between training and injection. Those animals that received posttraining saline injections showed a rather consistent and predictably high incidence of conditioned response retention. On the other hand, butyric acid-treated mice showed a temporal gradient for retrograde amnesia, i.e., the shorter the interval between training and butyric acid injection, the greater the incidence of retrograde amnesia, as measured with the testing trial. The greatest incidence of retrograde amnesia (80%) occurred among those mice given butyric acid immediately following training (0). By 4 min following training, although a significant incidence of retrograde amnesia was still evident (60%;

$p<0.05$), the amnesic effect of butyric acid appeared somewhat reduced. The amnesic effect of butyric acid treatment given 16 min post-training (20%) was no longer statistically significant ($p>0.10$), and there was no evidence for retrograde amnesia when butyric acid was given 64 min after training.

Several further observations emerge from the data. It is apparent that no effect of butyric acid is manifest at 23–24 hr following its intra-peritoneal administration. It is further apparent that the performance of the conditioned avoidance response was not affected by butyric acid when it was given 16 or more minutes following a training trial. When administered within approximately 4 min after a training trial, butyric acid did produce a retrograde amnesia, the incidence of which was time linked to the interval between training and its administration. The foregoing experiment supports the feasibility of using butyric acid as an amnesic agent and indicates the validity of its use in subsequent experiments.

M.   BUTYRIC ACID: EFFECTS UPON BRAIN 5-HT

In order to assess the effects of butyric acid administration upon brain 5-HT level, an experiment that paralleled the previously described behavioral study was carried out. Male CF-1S mice were given an intraperitoneal injection of 22.5 μmoles/g of butyric acid at pH 7.0 and were killed either immediately after injection (0) or 4, 16, or 64 min later. Tissues from the cerebral cortex, diencephalon, limbic region, and cerebellar cortex were removed and assayed, as described above, for their 5-HT contents (see Fig. 18).

By 16 min after butyric acid injection, or approximately 8 min following the onset of coma, 5-HT concentration was significantly elevated in the cerebral cortex, diencephalon, and limbic region. By 64 min following treatment, or just prior to recovery from the comatose state, brain 5-HT levels in these three areas as well as in the cerebellar cortex were still further elevated.

The significant increments in brain 5-HT concentration following butyric acid injection are consistent with observations that have been made in previous studies (Essman, unpublished data). Although the precise reason for this elevation has not been determined, several hypotheses may be offered. Increased 5-HT levels at the nerve ending could be a result of the direct central effects of butyric acid or secondary to the liberation of 5-HT by influx of other amines. There is evidence, as cited above, that hepatic coma is associated with an increase in

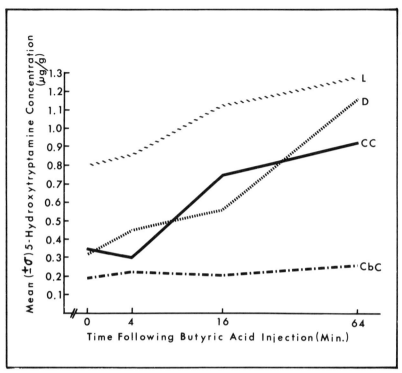

18   Regional 5-HT concentration as a function of time following intraperitoneal injection of butyric acid.

brain tryptophan, which could provide for increased brain 5-HT. The increase in non-albumin bound tryptophan observed in hepatic coma (Young et al., 1975) is consistent with the suggestion that brain tryptophan level is controlled by free plasma tryptophan (Tagliamonte et al., 1971). Another possibility is that 5-HT is elevated by a butyric acid-mediated reduction in monoamine oxidase (MAO) activity, as has been noted with tryptophan-induced hepatic encephalopathy (Ogihara et al., 1966). There does not seem to be a consistent relationship between either the onset or duration of butyric acid-induced coma and the elevation of brain 5-HT; explicitly, brain 5-HT levels were notably higher within the 5 min prior to recovery from coma than they were during deep coma from which the mice did not recover for at least 50 or more minutes. Thus, as we have pointed out earlier, the elevation in brain 5-HT necessary to cause a retrograde amnesia must not necessarily be excessive nor must it bear a specific relationship to the overt effects of the treatment which increased 5-HT levels. Clearly, in the present circumstances, a shorter duration coma may be as adequate an amnesic event; perhaps the only requirement is that 5-HT should be elevated at a point in time when the central events associated with the consolidation of the memory trace can be modified or interrupted.

It may be useful to remember that at least one form of seizure episode–myoclonus–follows butyric acid injection and precedes coma by a few minutes. It has been noted (Scherrer, 1968) that myoclonus occurring during phasic paradoxical sleep corresponds temporarily to the release of 5-HT. A role of 5-HT has been suggested in the etiology of human myoclonic movement disorders (Klawans et al., 1975). Myoclonus, induced in guinea pigs by 5-hydroxytryptophan, was antagonized by methylsergide; this same agent reduced myoclonus in neurological patients (Bedard and Bouchard, 1974; Goetz and Klawans, 1974). In the present study, it is of interest to consider whether the myoclonic phase of the butyric acid effect is the result of elevated brain 5-HT, or if elevated 5-HT is the result of the myoclonic episode. In either case, butyric acid-induced myoclonus in mice does appear to be related to the elevation in brain 5-HT.

N. BUTYRIC ACID: EFFECTS UPON CEREBRAL PROTEIN SYNTHESIS

An issue which the amnesic effects of butyric acid and its elevation of brain 5-HT are further concerned is the extent to which a change in brain protein synthesis is consistent with these central events. To investigate this premise, an *in vivo* study of protein synthesis was performed. It proceeded as follows: Intraventricular ^{14}C-leucine and intraperitoneal of 22.5 μmoles/g of butyric acid were administered to male CF-1S mice. Leucine injection was given either simultaneously with the butyric acid injection (0) or 4, 16, or 64 min later. A 5-min labeling pulse was permitted, the animals were killed, and the tissue was prepared as previously described for determination of the specific activities of leucine incorporated into proteins of the cerebral cortex, basal ganglia-plus-diencephalon, midbrain, and cerebellar cortex. Figure 19 shows ^{14}C-leucine incorporation into protein as a function of the time following butyric acid injection. By 16 min after butyric acid injection (or at a time when all animals had been comatose for approximately

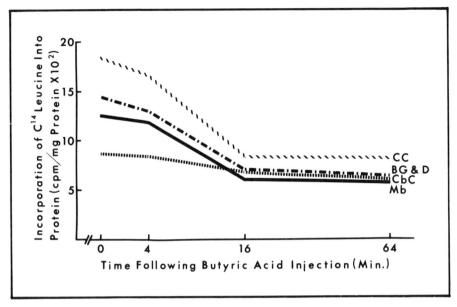

19 ^{14}C-leucine incorporation into regional brain proteins as a function of time following butyric acid injection.

8 min), a significant reduction in protein synthesis was observed for all regions of the brain, with the possible exception of the cerebral cortex. By 64 min, there was a 50% or greater reduction of amino acid incorporation into all regions concerned, except for cerebellar cortex. The data further support the general hypothesis that an amnesic agent capable of elevating brain 5-HT is also capable of mediating a reduction in regional protein synthesis. It is also noteworthy that the peak of 5-HT elevation occurred 16 min after butyric acid injection; this was also the point in time when the greatest decrease in protein synthesis was observed for this same brain region.

Utilizing the butyric acid model developed, two final series of experiments were undertaken. These studies utilized uric acid, previously shown to reduce the efficacy of an amnesic event and also to modify or reverse several of the biochemical changes associated with that event. The interaction of uric acid with (1) butyric acid-induced retrograde amnesia and (2) elevated regional 5-HT concentration and (3) reduced regional protein synthesis was studied in the following experiments.

Table I. Median Response Latency (Training and Testing) and
Incidence of Conditioned Avoidance Response Retention in Mice
Pretreated with Saline or Uric Acid Given
Posttraining Saline or Butyric Acid

Condition		Conditioned avoidance response retention		
Pretreatment	Posttraining	Median response latency		% CR retention
0.9% NaCl	0.9% NaCl	10.2	86.7	100
0.9% NaCl	Butyric acid	7.9	61.5	20[a]
Uric Acid	0.9% NaCl	9.4	85.5	100
Uric Acid	Butyric acid	8.6	83.5	70[a]

[a] $\chi^2 = 13.60$; $P < 0.005$.

O. BUTYRIC ACID-INDUCED RETROGRADE AMNESIA: EFFECTS OF URIC ACID

Groups of male CF-1S mice were given an intraperitoneal injection of either 0.9% NaCl or 2 mg/kg of uric acid dissolved in physiological saline. Sixty minutes later, all animals were given a single training trial in the apparatus previously described, providing for the acquisition of a passive avoidance response. Immediately following entry into the chamber in which footshock was delivered, groups of mice were given an intraperitoneal injection of either 0.9% NaCl or 22.5 μmoles/g of butyric acid. The design consisted of one NaCl- and one uric acid-pretreated group of mice given posttraining NaCl and one NaCl- and uric acid-pretreated group given posttraining butyric acid. The response latency for entry into the conditioning chamber on the training trial and for departure from the vestibule on the testing trial were recorded for each animal. Those animals achieving a 90-sec criterion of nonresponse were removed from the apparatus. A response latency of 12 sec or less on the training trial was taken as an indication of amnesia for the avoidance response, whereas a latency of 60 sec or more on the testing trial was suggestive of conditioned avoidance retention. All animals were given a single retention trial 24 hr following training at which time there was no evidence of residual effects of butyric acid treatment. The characteristics of the butyric acid-induced loss of righting reflex, myoclonus, and coma approximated that described earlier in this text.

The results of these experiments are shown in Table I; median response latencies for training and testing trials have been indicated, as has the incidence of conditioned response retention. Butyric acid administered following the training trial, requiring approximately 8 min for coma, produced a retrograde amnesia in at least 80% of the animals tested; this may be contrasted with the effect of posttraining saline administration where conditioned response retention was not modified. Butyric acid given to mice pretreated with uric acid 60 min earlier showed a significantly reduced amnesic effect. In this instance, only 30% of the animals tested showed a retrograde amnesia. These results strongly support earlier evidence that experimentally induced retrograde amnesia may be significantly attenuated by uric acid (Essman, 1970c). Furthermore, it is apparent that uric acid does not modify the already high incidence of conditioned response retention. These findings suggest that the effects of uric acid are probably operative on the mechanism or mechanisms by which retrograde amnesia is mediated.

P. BUTRYIC ACID-URIC ACID INTERACTIONS: 5-HT CONTENT AND PROTEIN SYNTHESIS

On the basis of previous observations, it would appear that at least one and possibly more of the effects of butyric acid are related to the alteration in cerebral 5-HT metabolism and its direct consequence, an inhibition of cerebral protein synthesis. It was the purpose of subsequent experiments to determine the extent to which butyric acid-mediated effects on brain 5-HT content and on protein synthesis would be affected by treatment with uric acid.

Pursuing this problem, male CF-1S mice were given an intraperitoneal injection of 0.9% NaCl or 2 mg/kg of uric acid. Sixty minutes later, the animals were given an intraperitoneal injection of butyric acid (22.5 μmoles/g), resulting in loss of righting reflex, brief episodes of phasic myoclonus, and onset of coma within approximately 8 min. Sixteen minutes following butyric acid injection, these animals were killed, their brain tissues excised and separated into regions consisting of cerebral cortex, diencephalon, limbic area, and cerebellum. The tissue was split so that the material constituting each half of each brain region was assayed for brain 5-HT concentration. The other half was incubated in an *in vitro* system similar to that previously described, to which ^{14}C-leucine was added. Under these conditions, the incorporation of ^{14}C into regional tissue proteins was measured.

The results of these experiments are shown in Table II. Several observations may be derived from these data; it is manifestly clear that uric acid prevented butyric acid-induced increases in regional 5-HT concentration. The differences between saline- and uric acid-

Table II. Mean ($\pm \sigma$) Regional 5-Hydroxytryptamine Concentration and
C^{14}-Leucine Incorporation into Protein Measured in Saline-
or Uric Acid-Treated Mice 60 Min following Butyric Acid Injection

Brain region	5-HT concentration		C^{14}-Leucine incorporation	
	NaCl	Uric acid	NaCl	Uric acid
Cerebral cortex	0.92	0.30[a]	820	1660[a]
	(0.18)	(0.07)		
Diencephalon	1.26	0.45[a]	640	1330[b]
	(0.27)	(0.11)		
Limbic area	1.28	0.86[b]	865	1180[b]
	(0.12)	(0.09)		
Cerebellum	0.26	0.22	600	840
	(0.04)	(0.04)		

[a] $P < 0.01$.
[b] $P < 0.05$.

pretreatment conditions reached statistical significance for all brain areas investigated, with the exception of the cerebellar cortex. In addition, uric acid given prior to the injection of butyric acid also prevented the butyric acid-induced decrease in ^{14}C-leucine incorporation into cerebral proteins. Again, these changes were statistically significant for all brain regions except for the cerebellar cortex.

Q. SOME CONCLUDING OBSERVATIONS

The results of the present experiments reinforce antecedant observations that the metabolic sequelae of an amnesic agent can be blocked, reversed, or attenuated by uric acid. Further, the amnesic potential of such memory-disruptive agents, such as ECS or butyric acid, in such circumstances, has been shown to be markedly reduced. While this observation does not apply universally to all agents or events effective in bringing about a retrograde amnesia in experimental animals, it is consistent with previous observations in which a number of events have been implicated in the elevation of brain 5-HT, alteration of 5-HT turnover, and/or inhibition of cerebral protein synthesis (Essman, 1973).

In an effort to explain the amnesic effect of ECS and its metabolic sequelae in terms of its alteration in oxidative metabolism and possible transient hypoxia, we have considered this possibility very carefully. It is our judgment, however, that such changes resulting from ECS that pertain to either memory consolidation or to macromolecular synthesis. Specifically, with regard to short chain fatty acid-induced coma, energy utilization and substrates of energy metabolism are not altered in cerebral cortex or brainstem of rats (Walker et al., 1970). This finding further supports the view that coma induced by butyric acid does not represent the result of altered cerebral energy metabolism and moreover, that butyric acid-induced amnesic and metabolic effects are relatively independent of oxidative metabolic changes.

The effects of uric acid upon amnesic production and the neurochemical effects of amnesic stimuli underscore a major point of our working hypothesis, i.e., that attenuation of a retrograde amnesic effect by a specific CNS-active agent (uric acid, in the present case) is consistent with the antagonistic effect of this same agent upon neurochemical changes produced by such an amnesic stimulus. Additional evidence demonstrating the effects of *in vitro* studies following *in vivo* events as well as *in vivo* effects for these same events supports the premise that the consolidation of the memory trace is an event to which alterations in specific metabolic pathways are linked. At least two such metabolic events are relevant: (a) changes in concentration, availability, or turnover on 5-HT; and (b) the synthesis of site-specific proteins. The cellular specificity of the latter event is dramatically illustrated by the observation that changes at the cellular level are reciprocal, i.e., at least one amnesic stimulus, ECS, decreases protein synthesis in neuronal perikarya and increases protein synthesis in adjacent glia. Likewise, reversal of an amnesic effect by uric acid also reverses the effect of the amnesic stimulus upon protein synthesis at the cellular level. Certainly, such cellular specificity bears further investigation as does the documentation of other specific physiological or pharmacological events that may be primarily or secondarily implicated in amnesic properties. It has been the purpose of this paper to explore some of the conditions and interrelationships between amnesia producing stimuli and their biochemical correlates, and to delineate in detail, more of the methods employed in that effort. It is our hope that in doing so, we have clarified the mechanisms that link the induction as well as reversibility of retrograde amnesia to the induction as well as reversibility of biochemical events in the CNS.

R. REFERENCES

Agranoff, B. W. (1969). Macromolecules and brain function. In: *Progress in Molecular and Subcellular Biology* (F. E. Hahn, ed.), pp. 203–212, Springer-Verlag, New York.

Baldessarini, R. J., and Fischer, J. E. (1973). Serotonin metabolism in rat brain after surgical diversion of the portal venus circulation. *Nature (London) New Biol.* 245, 25–27.

Barondes, S. H. (1970). Cerebral protein synthesis inhibitors block long-term memory. *Int. Rev. Neurobiol.* 12, 177–205.

Bedard, P., and Bouchard, R. (1974). Dramatic effect of methysergide on myoclonus. *Lancet* 1, 738.

Curzon, G., Kantamaneni, B. D., Winch, J., Rojas-Bueno, A., Murray-Lyon, I. M., and Williams, R. (1973). Plasma and brain tryptophan changes in experimental acute hepatic failure. *J. Neurochem.* 21, 137–145.

Cusumano, J. (1975). Experimentally induced coma in mice: A preliminary dose-response study with N-butyric acid. *Nucleus* 13, 20–22.

Essman, W. B. (1966). Effect of tricyanoaminepropene on the amnesic effect of electro-convulsive shock. *Psychoparmacologia* 9, 426–433.

Essman, W. B., and Essman, S. G. (1969). Enhanced memory consolidation with drug-induced regional changes in brain RNA and serotonin metabolism. *Pharmako-Psychiat. Neuropsychopharm.* 2, 28–34.

Essman, W. B. (1970a). Some neurochemical correlates of altered memory consolidation. *Trans. N. Y. Acad. Sci.* 32, 948–973.

Essman, W. B. (1970b). The role of biogenic amines in memory consolidation. In: *The Biology of Memory* (G. Adam, ed.), pp. 213–238, Akadamiai Kiado Publ., Budapest.

Essman, W. B. (1970c). Purine metabolites in memory consolidation. In: *Molecular Approaches to Learning and Memory* (W. Byrne, ed.), pp. 307–323, Academic Press, New York.

Essman, W. B. (1971). Drug effects and learning and memory processes. In: *Advances in Pharmacology and Chemotherapy* (S. Garattini and P. Shore, eds.), pp. 241–330, Academic Press, New York.

Essman, W. B. (1972a). Retrograde amnesia and cerebral protein synthesis: Initiation and inhibition by 5-hydroxytryptamine. *Totus Homo* 4, 61–67.

Essman, W. B. (1972b). Neurochemical changes in ECS and ECT. *Sem. Psychiat.* 4, 67–79.

Essman, W. B. (1973). *Neurochemistry of Cerebral Electroshock*, Spectrum Publ., New York.

Essman, W. B. (1974). Brain 5-hydroxytryptamine and memory consolidation. In: *Advances in Biochemical Psychopharmacology* (E. Costa, G. L. Gessa, and M. Sandler, eds.), Vol. 11, pp. 265–274, Raven Press, New York.

Essman, W. B. (1975). Diurnal differences in altered brain 5-hydroxytryptamine-related regional protein synthesis. *J. Pharmacol.* (Paris), 6, 313–322.

Goetz, C. G., and Klawans, H. L. (1974). Myoclonus, methysergide and serotonin. *Lancet* 1, 1284–1285.

Klawans, H. L., Goetz, C. G., and Bergen, D. (1975). Levodopa-induced myoclonus. *Arch. Neurol.* 32, 331–334.

Knell, A. J., Davidson, A. R., Williams, R., Kantamaneni, B. D., and Curzon, G. (1974). Dopamine and serotonin metabolism in hepatic encephalopathy. *Brit. Med. J.* 1, 549–551.

Lal, S., Aronoff, A., Garelis, E., Sourkes, T. L., Young, S. N., and de la Vega, C. E. (1975). Cerebrospinal fluid, homovanillic acid, 5-hydroxy-indolacetic acid and pH before and after probenecid in hepatic coma. *Clin. Neurosurg.* 22, 142–154.

Muto, Y. Y., Takahashi, Y., and Kawanura, H. (1964). Effect of short-chain fatty acid amines on the electrical action of neo-palio and archicortical systems. *Brain Nerve (Tokyo)* 16, 608.

Muto, Y. Y., and Takahashi, Y. (1965). Cited in: Medicine from abroad. *Post Grad. Med.* 37, A158.

Ogihara, K., Mozai, T., and Hirai, S. (1966). Tryptophan as cause of hepatic coma. *New Engl. J. Med.* 275, 1255–1256.

Rizzoli, A. A., and Galzinga, L. (1970). Molecular mechanism of the unconscious state induced by butyrate. *Biochem. Pharmacol.* 19, 2727–2736.

Samson, F. E., Dahl, N., and Dahl, D. R. (1956). A study on the narcotic action of the short-chain fatty acid. *J. Clin. Invest.* 35, 1291.

Scherrer, J. (1968). Les myoclonies physiologiques de la fase paradoxale du sommeil. *Rev. Neurol.* 119, 131–133.

Tagliamonte, A., Biggio, G., and Gessa, G. L. (1971). Possible role of "free" plasma tryptophan in controlling brain tryptophan concentrations. *Riv. Farmacol. Terap.* 2, 251–255.

Tozer, T. N., Neff, N. H., and Brodie, B. B. (1966). Application of steady state kinetics to the synthesis rate and turnover time of serononin in brain normal and reserpine treated rats. *J. Pharmacol. Exp. Therapeut.* 153, 177–182.

Walker, D. O., and Schenker, S. (1970). Pathogenesis of hepatic encephalopathy. *Amer. J. Clin. Nutrition* 23, 619–632.

Walker, D. O., McCandless, D. W., McGarry, J. D., and Schenker, S. (1970). Cerebral energy metabolism in short-chain fatty acid induced coma. *J. Lab. Clin. Med.* 76, 569–583.

Welch, A. S., and Welch, B. L. (1969). Solvent extraction method for simultaneous determination of norepinephrine, dopamine, serotonin, and 5-hydroxyindoleacetic acid in a single mouse brain. *Anal. Biochem.* 30, 161–179.

White, R. P., and Samson, F. E., Jr. (1956). Effects of fatty acid anions on the electroencephalogram of unanesthetized rabbits. *Amer. J. Physiol.* 186, 271.

Whittaker, V. P. (1969). The synaptasome. In: *Handbook of Neurochemistry* (A. Lajtha, ed.), Vol 2, pp. 327–364, Plenum Press, New York.

Young, S. N., Lal, S., Sourkes, T. L., Feldmuller, F., Aronoff, A., and Martin, J. B. (1975). Relationships between tryptophan in serum and cerebrospinal fluid, and 5-hydroxyindoleacetic acid in CSF of man: Effect of sclerosis of liver and probenecid administration. *J. Neurol. Neurosurg. Psychiat.* 38, 322–330.

Zieve, L. (1966). Pathogenesis of hepatic coma. *Arch. Int. Med.* 118, 211–223.

4

Neurochemical Correlates of Learning Ability

Y. TSUKADA
M. NOMURA
K. NAGAI
S. KOHSAKA
H. KAWAHATA
M. ITO

Department of Physiology
Keio Unversity School of Medicine
Shinanomachi, Tokyo, Japan

T. MATSUTANI

Department of Developmental Physiology
Institute for Comprehensive Medical Sciences
Fujita Gakuen University
Toyoake, Aichi, Japan

A. INTRODUCTION

This study was aimed at clarifying cerebral molecular mechanisms involved in higher nervous activities. Since it is well known that certain inborn errors of metabolism (e.g., phenylketonuria, cretinism) are frequently accompanied by severe mental retardation, it was assumed that metabolic disorders that occur during early stages of development cause effects on brain function, and especially on mental activity. Therefore, we attempted to produce experimental model animals for metabolic disorders such as cretinism, phenylketonuria, microencephaly, and excessive secretion of growth hormone and have examined both neurochemical changes and learning ability in these animals. Some neurochemical events involved in the control of learning ability are discussed below.

B. MATERIALS AND METHODS

1. Experimental Animals

Rats of the Wistar strain were used. To produce cretinous hypo-thyroid rats, 100 μCi of ^{131}I were injected into neonatal rats (Maloof et al., 1952). Histological observations indicated that this treatment produced an almost complete disappearance of the thyroid glands of these rats. Phenylketonuria was produced by subjecting rats to a diet containing high L-phenylalanine from the neonatal stage for several months (Tsukada, 1966). After delivery, mother rats were fed a diet containing 3% L-phenylalanine and their young were fed their mothers' milk which contained high phenylalanine. Thereafter, weanling rats were kept on a diet containing 7% L-phenylalanine for 4 months. Urinary excretion of phenylpyruvic acid and serum phenylalanine levels were chemically determined every week. Microencephalic rats were produced by injecting 20 mg/kg (i.p.) methylazoxymethanol (MAM) to pregnant rats on the fifteenth gestation day (Spatz and Laqueur, 1968). The off-spring became microencephalic, but otherwise their development was normal. Growth hormone treatment was carried out using daily injec-tions of bovine somatotropin (2 mg/kg) to developing rats from neo-natal stage for 3 months.

2. Chemical Determinations

After decapitation, brain tissue was excised in the cold and separated into several regions (e.g., cerebral hemisphere, brain stem, cerebellum, spinal cord). Brain tissue was fractionated by the method of Schmidt–Thannhauser, and its contents of DNA, RNA, phospholipid-P, and pro-tein were determined by conventional methods. The concentrations of free amino acids in ethanol–chloroform extracts of brain regions were measured by automated amino acid analysis (Nihon-Densi, JLC-6AH). Catecholamine contents were determined by a fluorometric method (Chang, 1964; Cox and Perhach, 1973), and indoleamine and its deriva-tives were determined by the method of Fischer et al. (1970). The activity of 2',3'-cyclic nucleotide 3'-phosphohydrolase (CNP) was assayed by the method of Kurihara and Tsukada (1967). The activities of thymidine kinase (Klemperer and Haynes, 1968) and DNA-dependent DNA polymerase (Bollum, 1968) in developing brain were assayed by conventional methods using brain homogenates.

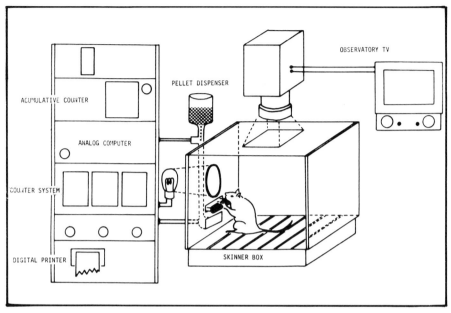

1   Discriminative learning apparatus.

3. Learning Test

At the age of 4 months, male rats were trained to discriminate the brightness of a light stimulus. A schematic representation of the experimental apparatus that was used is shown in Fig. 1. After preliminary training for shaping at a bar-pressing response, training for brightness discrimination was introduced. A bright light (2.5×10^4 fl) was used as the positive stimulus (S^+), and a dim light (0.5×10^4 fl) was used as the negative one (S^-). Whenever a rat pressed the bar under the bright light, it received a pellet, whereas pellets could not be obtained by pressing the bar when dim light was on. Each daily session consisted of 10 S^+ and 10 S^- periods. Each 20-sec stimulus was separated by a 5-sec blackout period and randomized in accord with the Gellerman series (Table I). Learning performance scores were recorded by counting the number of correct responses (R^+) and the number of errors (R^-), which provided the "correct response ratio" (R^+/R^++R^-). The criterion established for discrimination was three successive days of 85% "correct response ratio." Discrimination training was continued

Table I. Program for Operant Brightness Discrimination[a]

P - d - N - d - P - d - P - d - N - d - N - d - P - d - N - d - N - d - P - d - N - d - P - d - N -
d - N - d - P - d - P - d - N - d - P - d - P - d - N

P :   20 sec 2.5 x 10^4 fl, positive discriminative stimulus
N: 20 sec 0.5 x 10^4 fl, negative discriminative stimulus
d :   5 sec darkness
S^+:   numbers of responses by P
S^-:   numbers of responses by N

[a] During pretraining all rats were magazine trained, shaped to the bar press.

until either this criterion of discrimination or 35 daily sessions was attained.

C.   RESULTS AND DISCUSSION

1.   Hypothyroid Rat

In experimentally produced hypothyroid rats, body weight gain was clearly suppressed, their body weights at 4 months being less than half that of the control group (Fig. 2). Although brain weight was reduced by about 10%, the content of DNA per brain exhibited a normal value (Fig. 3). The contents of glutamate and aspartate in the brain tended to decrease, as compared to normal control animals, but the content of GABA remained almost normal (Fig. 4). On the other hand, the contents of cerebral dopamine, norepinephrine, and serotonin tended to increase (Table II). Phospholipid content in the brain is known to increase during the myelination process. In hypothyroid rats, cerebral phospholipid content was lower during the course of development than in the control (Fig. 5). Also, the activity of CNP, which is thought to be a good marker enzyme for myelin sheath (Kurihara and Tsukada, 1968), was lower than that of the control (Figs. 6 and 7).

From these results, it is postulated that myelination was inhibited considerably in the hypothyroid rat as has been previously shown by the histological observations of Eayrs (1966). Consequently, it is assumed that thyroid hormone affects not only somatic development, but also glial cell differentiation which in turn plays an important role in myelination of the CNS.

The learning ability of hypothyroid rats was examined under

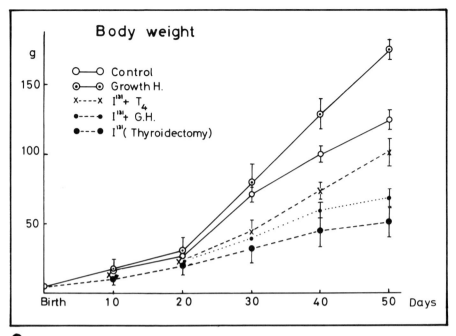

Body weight

g

Control
Growth H.
$I^{131} + T_4$
$I^{131} + G.H.$
I^{131} (Thyroidectomy)

150

100

50

Birth 10 20 30 40 50 Days

2

3

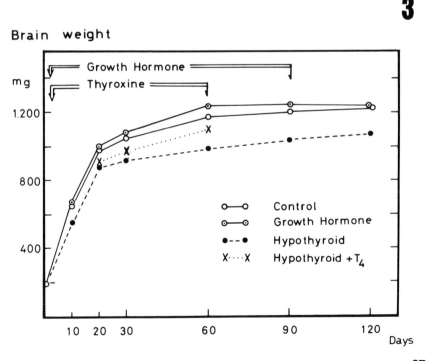

Brain weight

mg

Growth Hormone

Thyroxine

1.200

800

400

Control
Growth Hormone
Hypothyroid
X····X Hypothyroid + T_4

10 20 30 60 90 120

Days

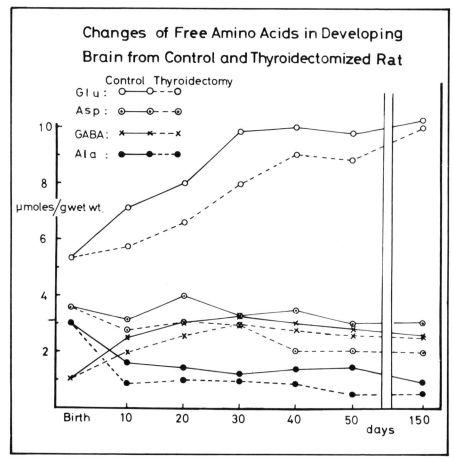

4   Changes of free amino acids in developing
brain from control and thyroidectomized rat.

brightness discrimination at the age of 4 months. Normal rats usually
reached the 85% level of "correct response ratio" within 25 sessions,
whereas hypothyroid rats exhibited such a dramatic loss of learning
ability that they did not achieve even the 70% level after 35 sessions
(Fig. 8). With the discrimination learning test, hypothyroid rats showed
very poor performances although previously reported behavioral data
on hypothyroid rat has been unequivocal (Denenberg and Myers, 1958;
Eayrs, 1966). Therefore, it is considered that the poor learning performance that accompanies experimentally produced hypothyroidism
in the rat might correspond to the behavior of cretinism.

Table II. Monoamine Contents in Rat Brain

5 Months old	Control				Hypothyroid			
	½ Brain (mg wet wt)	DA	NA	5-HT (µg/g wet wt)	½ Brain (mg wet wt)	DA	NA	5-HT (µg/g wet wt)
Cerebral hemisphere	285 (4)	0.40 ± 0.11 (4)	0.30 ± 0.12 (4)	1.11 ± 0.09 (4)	259 (4)	0.80 ± 0.12 (4)	0.47 ± 0.05 (4)	1.33 ± 0.12 (4)
Brain stem[a]	224 (4)	0.94 ± 0.13 (4)	0.60 ± 0.07 (4)	1.37 ± 0.14 (4)	179 (4)	1.09 ± 0.31 (4)	0.75 ± 0.12 (4)	1.57 ± 0.12 (4)
Cerebellum	115 (4)	0.14 ± 0.06 (4)	0.18 ± 0.04 (4)	—	90 (4)	0.24 ± 0.09 (4)	0.28 ± 0.09 (4)	—

[a] Diencephalon, mesencephalon, and pons.

69

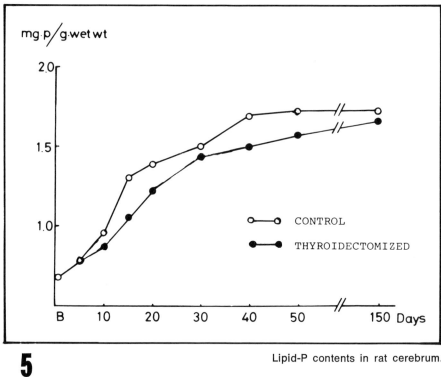

5

Lipid-P contents in rat cerebrum.

6

Adenosine 2'3'cyclic phosphate → Adenosine 2'phosphate

2'3'cyclic nucleotide 3'phosphohydrolase (CNP)

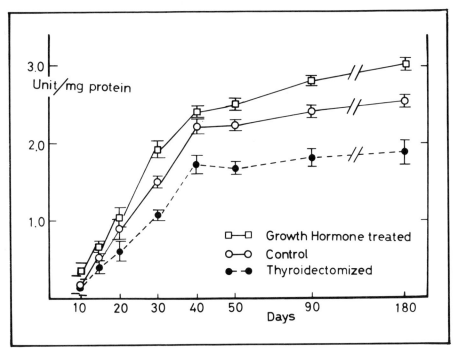

7

CNP activity in rat brain.

At 2–3 days after radiothyroidectomy, hypothyroid rats were treated by a replacement therapy. L-Thyroxine (T4; 1 μg/g) was injected daily for 2 months. Body weight, brain weight, and cerebral amino acid contents were restored to normal, and the activity of CNP was also moderately restored (Table III). However, the learning performances of rehabilitated rats were still very poor (Fig. 8).

These results indicated that continuous secretion of thyroid hormone may be essential for the development of normal brain function. If hypothyroid rats were rehabilitated by injection of bovine somatotropin instead of L-thyroxine, their learning performances were improved to a greater extent than those of thyroxine-treated cretinous rats (Fig. 8). In this case, the restoration of CNP activity was also greater than that of thyroxine-treated rat. Again, it was shown that cerebral CNP activity is closely related to learning ability.

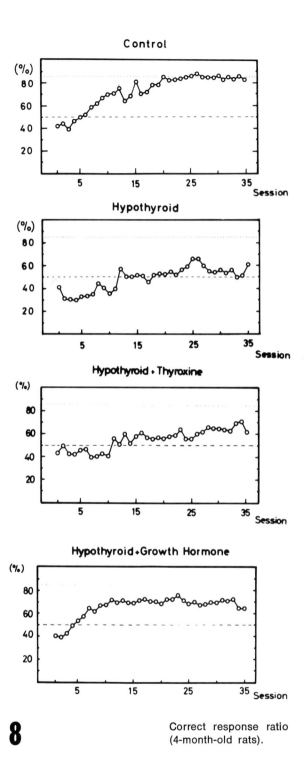

Control

Hypothyroid

Hypothyroid + Thyroxine

Hypothyroid + Growth Hormone

8   Correct response ratio
(4-month-old rats).

Table III. 2',3'-Cyclic Nucleotide
3'-Phosphohydrolase

Condition	Activity[a] (U/mg protein)
Control	1.51 ± 0.06 (24)
Hypothyroid (I^{131})	1.09 ± 0.10 (31)
I^{131} + T_4	1.20 ± 0.08 (4)
I^{131} + GH	1.50 ± 0.09 (4)
GH	1.89 ± 0.10 (5)

[a] Numbers in parentheses are number of 30-day-old rats.

2.   Phenylketonuric Rats

In experimentally produced phenylketonuric rats, gains of body weight and brain weight were considerably suppressed, whereas the content of DNA per brain did not differ from that of normal controls. This could be taken as evidence that the brain cell population had developed normally. Cerebral phenylalanine and tyrosine contents were clearly elevated, whereas that of serotonin was significantly decreased. However, catecholamine content was not changed. The cerebral CNP activity of 4-month-old phenylketonuric rats was significantly lower than that of the control. Hence, it is suggested that induction of a hyperphenylalanemic state during an early stage of development might affect glial cell function, particularly that of oligodendrocytes, such that myelination might be disturbed.

Neurochemical data on phenylketonuric rats are shown in Table IV. The learning performances of phenylketonuric rats was so poor that they could not achieve 80% level of "correct response ratio" even after 35 sessions (Fig. 9). Extinction also required a much longer period than with the controls. As another example, a study performed in our laboratory (Noguchi et al., 1974) revealed that when hydrocortisone acetate (40 mg/kg) was injected daily to neonatal rat for 4 days, DNA metabolism in the brain was disturbed, and CNP activity was also suppressed when rats attained the age of 4 months. The learning ability of these hydrocortisone-treated rats was also poor.

In the studies mentioned above, the developing infant brains apparently were affected functionally and the "attack period" corresponded to the stage of powerful glial proliferation (Fig. 16). Therefore, the myelination process could have been disturbed by an effect which

Table IV. Neurochemical Characteristics of Phenylketonuric Rats
(7 months old)

	Control	PKU[b]
Body weight (g)	270 ± 35 (5)	173 ± 19 (5)
Brain weight (mg)	1332 ± 65 (4)	1207 ± 58 (4)
DNA/brain (mg)	1.16 ± 0.06 (4)	1.13 ± 0.04 (4)
Glutamate (μmole/g)	9.54 ± 0.12 (3)	11.6 ± 0.52 (3)
Aspartate (μmole/g)	2.53 ± 0.61 (3)	2.55 ± 0.40 (3)
GABA (μmole/g)	2.82 ± 0.26 (3)	2.61 ± 0.25 (3)
Phenylalanine (μmole/g)	0.08 ± 0.02 (3)	0.36 ± 0.11 (3)
Tyrosine (μmole/g)	0.08 ± 0.02 (3)	0.22 ± 0.08 (3)
Serotonin[a] (μg/g)	0.75 ± 0.04 (3)	0.54 ± 0.07 (3)
Dopamine[a] (μg/g)	0.96 ± 0.27 (3)	0.86 ± 0.07 (3)
Norepinephrine[a] (μg/g)	0.30 ± 0.05 (3)	0.24 ± 0.06 (3)
CNP (U/mg protein)	3.16 ± 0.10 (5)	2.76 ± 0.13 (5)

[a] 4-month-old rats.
[b] Underlined values were statistically significant.

was reflected in the lower activity of CNP. At the same time, the learning performances of these animals were poor. Therefore, we assumed that cerebral CNP activity is a good biochemical indicator of learning ability.

3. Growth Hormone Treatment

In the next study, the effect of growth hormone on postnatal brain development was examined. Bovine somatotropin (2 mg/kg) was injected daily to neonatal rats for 3 months. The gain of body weight was higher than that of the controls, but only very slight increases of brain weight and brain DNA were observed. This indicated that somatotropin may not affect brain cell proliferation in the postnatal period (Figs. 2 and 3). However, cerebral CNP activity was clearly higher than that of controls during the course of development (Fig. 7). It seems that myelination in the cerebrum is accelerated by growth hormone treatment.

The learning ability of these rats was tested when they were 4 months old. Growth hormone-treated rats were able to learn rapidly, achieving the 85% level within 10 sessions (Fig. 10). From these results, it is suggested that growth hormone might enhance both myelination

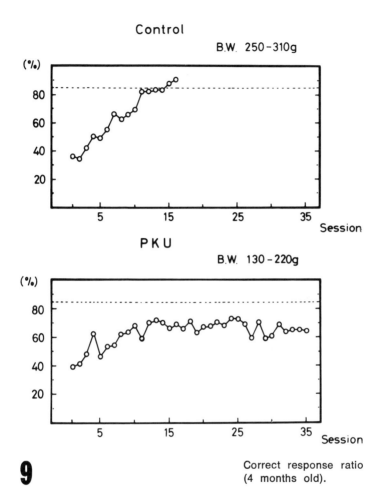

9

Correct response ratio
(4 months old).

and learning ability. A close relationship between cerebral CNP activity and learning ability was also found in these studies.

It has been reported (Block and Essman, 1965; Zamenhof et al., 1966) that when growth hormone, in smaller doses, was administered to the pregnant rat, the offspring displayed a significant improvement in extinction of a conditioned avoidance response at maturity. It was suggested that this effect of growth hormone might be attributed to an increase of the number of cortical neurons. In our experiment, in which growth hormone was injected postnatally, it seems more likely that

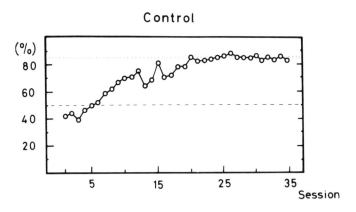

Control

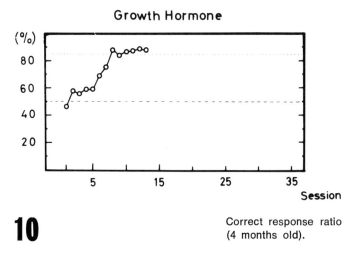

Growth Hormone

10   Correct response ratio
(4 months old).

its effect could have been exerted on glial cell proliferation or glial cell function.

4. Microencephalic Rats

By single injections of MAM to pregnant rats at the fifteenth gestation day, the increase of DNA in the fetal brain was strongly inhibited for 4 days. During this period, only neuronal cell proliferation was observed; glial cell proliferation had not yet begun. Therefore, cell damage by MAM treatment appeared only in neuroblasts or young

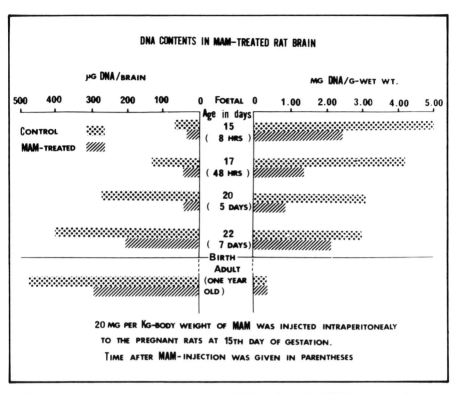

11   DNA contents in MAM-treated rat brain.

neurons. At 4 days after MAM injection, cerebral DNA began to be synthesized and its content increased during the course of development (Fig. 11). The fetuses were normally delivered from MAM-treated mother rats and their body weight gains were almost normal. However, the sizes of their cerebral hemispheres were very small and their weights were only about half those of control animals (Fig. 12). The content of DNA in the cerebrum was also decreased to about half that of the control. This indicated that a reduction of both neurons and glial cells had occurred.

At the age of 1 to 6 months, the CNP activity of the cerebrum appeared normal (Table V). However, the contents of monoamines (e.g., dopamine, serotonin) were clearly increased in the cerebral hemisphere (Figs. 13 and 14). On the other hand, the contents of GABA, glutamate, aspartate, and glycine in the brain were normal.

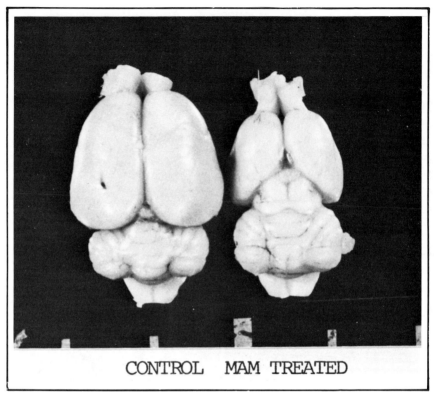

CONTROL   MAM TREATED

12

MAM-microencephaly (9-month-old rat).

Table V. 2′,3′-Cyclic Nucleotide 3′-Phosphohydrolase
in Developing Rat
Cerebral Hemisphere[a]

	Control (U/mg protein)	MAM (U/mg protein)
25 days	2.34	2.71
6 months	3.50	3.40

[a] Wistar-Imamichi strain.

Also, cerebral cholineacetyltransferase activity did not differ from that of the controls.

From the neurochemical data obtained from MAM-treated rats, it is suggested that myelination processes proceeded normally, but that monoamine-containing synapses showed abnormal development.

Test of learning performance in these animals indicated that their ability for discrimination learning was surprisingly good, but that they

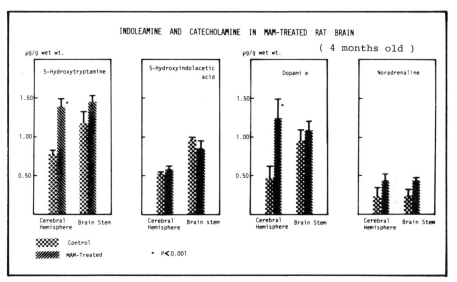

13   Indoleamine and catecholamine in MAM-treated rat brain (4 months old).

14   Developmental changes of 5-HT contents in cerebral hemisphere of MAM-treated rat. Twenty milligrams per kilogram body weight of MAM was injected intraperitoneally to pregnant rat at the 15th day of gestation.

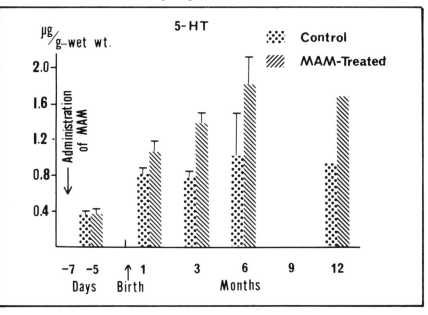

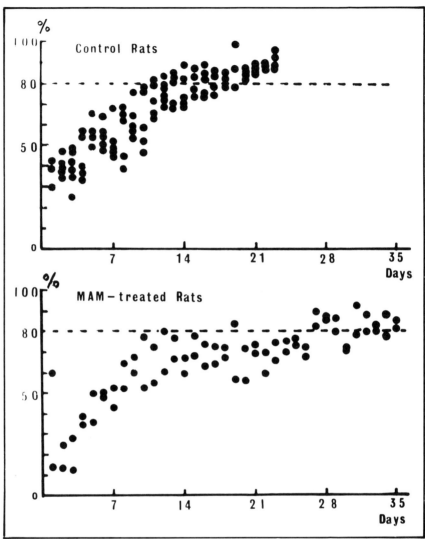

15   Correct response ratio (6-month-old male Wistar-Imamichi rats).

took a much longer time to reach the 85% level of "correct response ratio" (Fig. 15). MAM-treated rats behaved quite normally though their learning behavior was slightly disturbed. Hence, it is suggested that neuronal damage occurring in the prenatal stage could be functionally compensated, to a great extent, during the course of postnatal development so long as myelination proceeds normally. The physiological significance of the observed increases in monoamine contents of MAM-treated rats remains obscure.

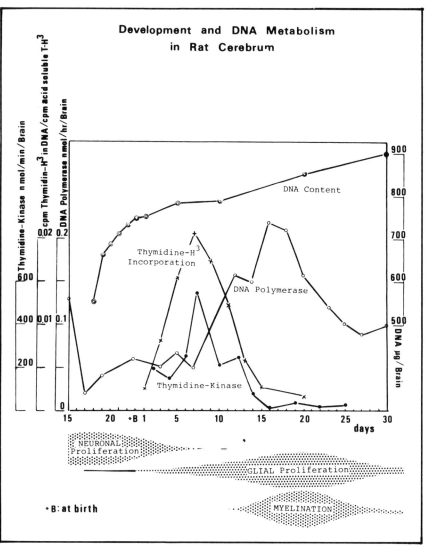

16   Development and DNA metabolism in rat cerebrum.

5.   DNA Synthesis in Developing Rat Brain

The enzymes concerned with DNA synthesis were assayed in developing rat brain. The content of DNA was sharply elevated during the late fetal stage and increased gradually during the postnatal period. On the seventh and sixteenth postnatal days, the respective peaks of thymidine-kinase and DNA-dependent DNA polymerase were obtained (Fig. 16). The appearance of peak activities of these enzymes might

Table VI. Relation between Cerebral Components and Learning Ability[a]

Experimental model for brain damages	Attack period	Cerebral weight	Cerebral DNA	Cerebral CNP	Learning ability	Neurotransmitter		Body weight
						Mono-amines	GABA	
Hypothyroid rat (Cretinism)	Neonatal, immediately after birth							
Experimental phenylketonuric rat (PKU)	Neonatal, 0 day–8th month							
Hydrocortisone-treated rat	Neonatal, 0–3rd day							
Growth hormone treated rat	Neonatal, 1st day–3rd month							
MAM-treated rat (Microcephalus)	Prenatal, 15th day of gestation							

[a] The determinations were carried out on 4-month-old rats. CNP: 2',3'-cyclic nucleotide 3'-phosphohydrolase. Learning ability: brightness discriminative learning.

   or : slightly decreased or increased
   or : moderately decreased or increased
       : markedly decreased
       : unchanged

correspond to glial cell proliferation, and some insults given during this period could disturb subsequent myelination processes.

D. SUMMARY

The correlation between neurochemical events and learning ability was investigated using hypothyroid, phenylketonuric, microencephalic, and growth hormone-treated rats (see Table VI). The results obtained showed that a clear relationship existed between discrimination learning ability and 2', 3'-cyclic nucleotide 3'-phosphohydrolase activity, which is a marker of myelination in the CNS. The cerebral contents of putative neurotransmitters were not so clearly related to learning ability. The importance of myelination in the development of higher nervous activities was emphasized. Neuronal cell defects occurring during the prenatal period did not cause a severe retardation in the development of learning ability so long as myelination proceeded normally.

E. REFERENCES

Block, J. B., and Essman, W. B. (1965). Growth hormone administration during pregnancy: A behavioral difference in offspring rats. *Nature (London)* 205, 1136–1137.

Bollum, F. J. (1968). Filter paper disk techniques for assaying radioactive macromolecules. *Methods in Enzymology*, Vol. 12B, pp. 168–173, Academic Press, New York.

Chang, C. C. (1964). A sensitive method for spectrophotofluorometric assay of catecholamines. *Int. J. Neuropharm.* 3, 643–649.

Cox, R. H., Jr., and Perhach, Jr. (1973). A sensitive, rapid and simple method for the simultaneous spectrophotofluorometric determinations of norepinephrine, dopamine, 5-hydroxytryptamine and 5-hydroxyindoleacetic acid in discrete areas of brain. *J. Neurochem.* 20, 1777–1778.

Denenberg, V. H., and Myers, R. D. (1958). Learning and hormone activity. I. Effects of thyroid levels upon the acquisition and extinction of an operant response. *J. Comp. Physiol. Psychol.* 51, 213–219.

Eayrs, J. T. (1966). Thyroid and central nervous development. Scientific Basis of Medicine. Ann. Rev. (London), Athlone Press, 317–339.

Fischer, C. A., Kariya, T., and Aprison, M. H. (1970). A comparison of the distribution of 5-hydroxyindoleacetic acid and 5-hydroxytryptamine in four specific brain areas of the rat and pigeon. *Comp. Gen. Pharm.* 1, 61–68.

Klemperer, H. G., and Haynes, G. R. (1968). Thymidine kinase in rat liver during development. *Biochem. J.* 108, 541–546.

Kurihara, T., and Tsukada, Y. (1967). The regional and subcellular distribution of 2′,3′-cyclic nucleotide 3′-phosphohydrolase in the central nervous system. *J. Neurochem.* 14, 1167–1174.

Kurihara, T., and Tsukada, Y. (1968). 2′,3′-Cyclic nucleotide 3′-phosphohydrolase in the developing chick brain and spinal cord. *J. Neurochem.* 15, 827–832.

Maloof, F., Dobyns, B. M., and Vikery, A. L. (1952). The effects of various doses of radioactive iodine on the function and structure of the thyroid of the rat. *Endocrinology* 50, 612–638.

Noguchi, T., Suda, H., and Tsukada, Y. (1974). Studies on DNA metabolism in developing rat brain with special reference to the effect of hydrocortisone. (in Japanese) *Shinkei Kagaku.* (*Bulletin of the Japanese Neurochemical Society*) 13, 190–193.

Spatz, M., and Laqueur, G. L. (1968). Transplacental chemical induction of microencephaly in two strains of rat. I. (33404). *Proc. Soc. Exp. Biol. Med.* 120, 705–710.

Tsukada, Y. (1966). Amino acid metabolism and its relation to brain functions. *Progr. Brain Res.* 21A, 268–291.

Zamenhof, S., Mosley, J., and Schuller, E. (1966). Stimulation of the proliferation of cortical neurons by prenatal treatment with growth hormone. *Science* 152, 1396–1397.

5

Alterations in the Structure and Function of Free and Membrane-Bound Polysomes and Messenger-RNA of Neuronal and Glial Enriched Fractions of Rat Cerebral Cortex During Light Deprivation and Exposure to Lights of Different Wavelengths[1]

M. R. V. MURTHY
HUGUETTE ROUX
A. D. BHARUCHA
R. CHARBONNEAU

Department of Biochemistry
Faculty of Medicine, Laval University
Québec, Canada

[1]This work was supported by research grants from the Medical Research Council of Canada and from the Ministry of Education, Government of Québec. We wish to express our thanks to Dr. G. Bernardi, Institut de Biologie Moléculaire, C.N.R.S., Paris, and to Dr. J. Daillie, Université Claude Bernard, Lyon, for the facilities accorded to one of the authors (M. R. V.) in their laboratories during the preparation of some parts of this manuscript. We record our appreciation to Mme Umamani Rao for her excellent technical assistance in different phases of this work.

A. INTRODUCTION

Environmental stimuli have long been known to have decisive effects on cerebral development, whether this is measured at the psychological, morphological, or molecular levels. Thus, in the absence of certain environmental demands, neuronal organizations that normally respond specifically to these environmental stimuli suffer atrophy, and as a result, changes are produced in the biochemical mechanisms regulating brain metabolism. On the other hand, when imbalanced sensory demands are imposed, abnormal neural growth patterns are observed, and when changes in the modalities of environmental stimuli occur, these are paralleled by changes in the structural and functional characteristics of the affected regions of the nervous system (Riesen, 1975).

Although the visual system is not unique in its susceptibility to alterations in the environment, a large majority of experiments on the effects of sensory deprivation on brain function have used this system as a model. This is partly due to the relative ease of manipulation of visual stimuli in terms of several physical parameters (e.g., frequency, illumination, temporal pattern). Another reason is that, in this system, the input is mediated through a single pathway, the optic tract, thus facilitating a comparatively rigorous control of the sensory stimulus.

Effects of light deprivation have been investigated using different experimental approaches, such as changing the frequency composition of the incident light, monocular or binocular eyelid suture, rearing the animals in a dark enclosure, and enucleation. Various structures of the visual system and some structures that are related directly or indirectly to its function (e.g., retina, optic tract, colliculus, lateral geniculate nucleus, visual cortex and other cortical areas, dendrites, dendritic spines, synaptic sites) have been examined with the view of localizing the effects of deprivation, resumption, or increased input of light (Globus, 1975). The results have indicated that reduced visual stimulation causes an overall disruption of morphological organization in all areas of the visual system. The severity and irreversible nature of these effects appear to be greater as the phylogenetic scale is ascended and are more pronounced when the treatment is begun in younger animals (Chow and Stewart, 1972; Hubel and Wiesel, 1970; Burke and Hayhow, 1968). Alterations in the quality and quantity of administration of light have been observed to produce changes in the metabolism of carbohydrates, proteins, nucleic acids, neurotransmitters, specific classes of enzymes, cofactors, and other molecules of importance to normal cell function (Walker et al., 1975). Among these, the synthesis and turnover of RNA and protein have received particular attention (Rasch et al., 1961; Appel et al., 1967; Singh and Talwar, 1967, 1969; Richardson and Rose, 1973a,b), partly in view of the importance of these macromolecules in the elaboration of structural organization in the nervous system in the course of differentiation and development, and partly as a result of the popular hypothesis that these molecules may be directly implicated in the behavioral and learning processes.

Ribosomes of eukaryotes, including those of brain cells, exist in at least two forms which are distinguishable biochemically and morphologically: the so-called free (f-) ribosomes and the membrane-bound (b-) ribosomes. The latter, which occur extensively in secretory cells, are generally thought to be responsible for the synthesis of proteins for export, whereas f-ribosomes, which are more abundant in rapidly growing embryonic cells and in dedifferentiated tumor cells, are perhaps involved in the formation of proteins for the internal needs of the cell. This demarcation does not appear to be as clear cut as originally believed since certain proteins are found to be synthesized equally well by both f- and b-ribosomes, for example, NADPH cytochrome c reductase (Ragnotti et al., 1969), catalase (Higashi et al., 1972), and globin (Morrison and Lingrel, 1975). However, these two types of ribosomes seem to be exclusively involved in the formation of certain specific proteins; e.g., serum albumin (Takagi and Ogata, 1971), immunoglobulins (Schechter, 1973), and glycoproteins (Hallinan et al., 1968) are formed by b-ribosomes, while ferritin (Puro and Richter, 1971) and myosin (Nihei, 1971) are synthesized by free polysomes. Recently, Zomzely-Neurath and colleagues (1973) have shown that the brain-specific proteins S-100 and 14-3-2 are preferentially synthesized on free polysomes. Richardson and Rose (1973a,b) found that enhancement of amino acid incorporation into various areas of retina and cerebral cortex caused by exposure to light was due to the formation of a certain number of unidentified specific proteins and that it did not represent a general increase in protein synthetic activity. Singh and Talwar (1969) have reported that S-100 protein was one of the proteins rapidly synthesized in the visual cortex in response to flickering light. A drastic reduction in the cytoplasmic ribonucleoprotein particles and an irregular dispersion of filiform Nissl substance were noted in the retina (Brattgard, 1952; Rasch et al., 1959). It has also been shown that differential increases in the total number of ribosomes or b-ribosomes occurred in different classes of cells of the hippocampus of rats following acquisition of brightness discrimination (Wenzel et al., 1975) or conditioned reflex (Tushmalova and Prokofieva, 1973). Taken together, these findings indicate that f- and b-ribosomes may each have a specific role in the synthesis of proteins during altered sensory stimulation, and may themselves be affected by such changes.

The brain is composed of two major classes of cells, neurons and glia. The specific function of glial cells in the brain is as yet unclear. Neurons, on the other hand, are believed to be the main cellular links in the complex network for transmission and processing of information in the nervous system. However, it has been observed that there occurs an increased proliferation of glial cells, particularly in the fibrous portion of the neocortex, when rats are exposed to enriched environments (Altman and Das, 1964; Diamond et al., 1966), indicating that changes in sensory input may affect these cells as well as the neurons.

Our objectives in undertaking the present work were: (1) to examine the effect of light deprivation and subsequent light exposure on the synthesis, structure, relative proportion, and protein synthesizing capacity of f- and b-ribosomes and on the base composition of mRNA associated with these ribosomes; (2) to compare the effects of these treatments on neurons and glia in the optical cortex and in the rest of the cerebral cortex and finally in a completely dissimilar nonneural tissue, the liver. Light deprivation experiments were carried out on rats immediately after birth since, as mentioned earlier, the effects of sensory deprivation have been shown to be most severe at this age of the animal.

B.   MATERIALS AND METHODS

The main steps of the experimental protocol used in this work are indicated in the flow diagram of Fig. 1.

1.   Light Deprivation and Light Exposure

Pregnant Sprague–Dawley rats were kept in a dark room 3–4 days prior to expected parturition. Immediately after birth, male rats were selected out from the mixed litters and were divided at random into four groups, each consisting of 6–9 newborn animals. All manipulations were carried out, as far as feasible, in the dark and in some cases under dim green light. Medical X-ray film was placed in the dark room during the manipulations and developed immediately afterward, in order to confirm that no light leakage had occurred. One of the four groups of rats (normal) was taken out of the dark room and reared for 4 weeks under a regulated 12 hr light/dark cycle (group N4). Light was provided by a combination of overhead fluorescent light and diffuse day light, the total intensity at the floor of the cage being approximately

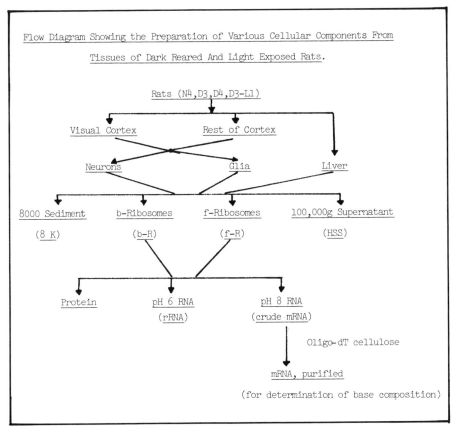

Flow Diagram Showing the Preparation of Various Cellular Components From Tissues of Dark Reared And Light Exposed Rats.

1 Flow diagram showing the preparation of various cellular components from tissues of dark-reared and light-exposed rats. The term "ribosomes" is used in this text to indicate a mixture of polysomal aggregates of different size classes which include a small proportion of dimers and monomers. as determined by sucrose density gradient (Murthy, 1972a).

250 ft cds during the light period and 3–5 ft cds during the dark period. The other groups of rats were maintained in the dark, two groups for a period of 3 weeks (groups D3) and one group for a period of 4 weeks (group D4). One group of the D3 rats was transferred to the light/dark cycle, as described for normal rats, following 3 weeks of dark rearing. They were kept in this cycle for 1 week before being sacrificed (group D3-L1). Each group of animals was maintained in separate communal cages with unlimited access to food and water and sufficient space for

free movement. Except for light deprivation, the experimental animals received all the usual environmental stimuli experienced by the normal animals, including handling by the personnel in charge. The ambient temperature varied from 23° to 25°C.

The apparatus used for producing the flickering light was constructed locally by the Laval University workshop, in such a manner that the frequency of flicker as well as light intensity could be regulated. For our present experiments, the flicker was set at 7/sec, since frequency in this range is reported to lead to maximum incorporation of amino acids into protein in the cerebral cortex of animals previously kept in darkness (Talwar et al., 1966; Singh and Talwar, 1967). The experimental animals were housed in black painted plastic boxes closed on all sides except for a small window on one wall which was fitted with a density filter or blue filter (range, 380–520 nm) or red filter (600 nm upward). The intensity of light on the side of the filter inside the cage was adjusted to 100 ft cds for all types of light exposure. The boxes containing the animals were ventilated by means of spiral tubes which permitted airflow without permitting any stray light into the boxes.

2. Preparation of Neurons and Glia

At the end of the experimental period, the rats were killed by decapitation and the tissues of interest were quickly removed. Tissue removal and the preparation of subcellular fractions were carried out at 0°–4°C. Tissues from two to three rats were pooled so that each group furnished three separate samples for analysis. Cell fractions enriched in neurons and glia were isolated from the cerebral cortex according to Rose (1967) by centrifugation through a discontinuous Ficoll-sucrose gradient.

3. Preparations of Subcellular Fractions

The 8000 g sediment (8K), membrane bound ribosomes (b-R), free ribosomes (f-R), and the 100,000 g supernatant (HSS) from brain were prepared according to previously published procedures (Murthy, 1972a), except for the modification that the f- and the b-ribosomes were not subjected to the second discontinuous density gradient centrifugation that was required in the original procedure. The structural properties, protein synthetic activities, and the RNA and protein constituents of these ribosomes have been characterized in detail in our previous papers (Murthy, 1970, 1972a,b). Using this method for preparation of b- and f-polysomes, Zomzely-Neurath et al., (1973) have recently demonstrated that the brain-specific proteins, S-100 and 14-3-2, are primarily syn-

thesized by free polysomes. The f- and b-ribosomes of liver were prepared according to Bloemendal et al. (1967).

4. Isolation of Proteins and RNAs

Ribosomal RNA (pH 6 RNA) and crude messenger RNA (pH 8
RNA) were isolated from f- and b-ribosomes by a modification (Murthy,
1972b) of the method of Hadjivassiliou and Brawerman (1967). After
successive extractions of pH 6 and pH 8 RNAs, the interphase layer,
the phenol layer, and the sediment were combined. Protein was isolated
from this following several extractions with ethanol:ether (4:1,v/v) to
remove phenol. The insoluble precipitate was washed twice with 5%
trichloracetic acid (TCA) and heated once at 90°C for 15 min with
5% TCA to degrade any contaminating nucleic acids. The residue was
washed again with ethanol and ether before determining radioactivity.

Messenger RNA (mRNA) was purified from the pH 8 RNA by
affinity chromatography on oligo-dT cellulose according to Aviv and
Leder (1972). Passage through oligo-dT cellulose columns was performed twice, since it has been reported that the first passage does not
completely remove the ribosomal RNAs (rRNAs) (Gorski et al., 1974).

5. Base Composition of Messenger RNA

The samples of mRNA recovered from the oligo-dT cellulose columns
were lyophilized and were digested with pancreatic RNase and RNase
T1 according to Darnell et al. (1971) in order to degrade the non-poly-A
portion of mRNA. The undegraded poly-A fragments were separated
from the digested nucleotides by affinity chromatography on oligo-dT
cellulose. The mRNA degradation products in the eluate were deionized
by passage through a Bio-Gel P2 column and they were further converted to the level of nucleosides by a second digestion with pancreatic
ribonuclease A, snake venom phosphodiesterase and alkaline phosphatase. Base composition was determined by tritium postlabeling of
these nucleosides according to Randerath and Randerath (1973). For
this, the nucleosides were oxidized with periodate followed by reduction
with tritiated potassium borohydride. The resulting labeled nucleoside
alcohols were separated by thin-layer chromatography. After identification of the radioactive regions by fluorography (Randerath, 1970), the
spots corresponding to the trialcohols of adenosine, guanosine, cytidine,
and uridine were cut out and radioactivies were determined to obtain
the relative base composition.

6. Radioactivity Measurements

A Mark 1 liquid scintillation counter (Nuclear Chicago) was used for all radioactivity measurements. Whenever simultaneous incorporations of ^{3}H-phenylalanine and ^{14}C-uridine were studied, radioactivities due to these two isotopes in a given sample were determined by differential counting in separate channels and disintegrations/minute (dpm) for each isotope were calculated using a previously established calibration curve. Protein and RNA samples were dissolved in 0.1 to 0.2 ml of NCS solubilizer before adding 15 ml of toluene-POPOP as scintillation fluid (40 ml spectrafluor mixed with one liter of redistilled toluene). Aqueous solutions were counted using Aquasol.

7. Other Analytical Procedures

DNA was determined according to Burton (1962) and RNA by the method of Fleck and Begg (1965) with the modification suggested by Sripati et al. (1967) for brain tissue. Protein was estimated by the method of Lowry et al. (1951).

C. RESULTS AND DISCUSSION

The RNA/DNA ratio decreased in both neurons and glia of the visual cortex when newborn rats were subjected to prolonged darkness immediately after birth (Table I). Four weeks of darkness had a greater detrimental effect than 3 weeks, and the neurons were more susceptible to this treatment than glia. Exposure of rats to 1 week of light/dark cycle following 3 weeks of darkness was not successful in restoring the RNA/DNA balance in either of these cell types. The liver cells, on the contrary, were unaffected by up to 4 weeks of dark rearing.

When the neurons and glia of the visual cortex of control rats (N4) were separated into different subcellular fractions, it was found that approximately one-fourth of the total RNA was present in the nuclear–mitochondrial–synaptosomal fraction (8K), half in the f-ribosomes, and the rest was divided about equally between the b-ribosomes and the HSS fraction. In control liver, the proportion of RNA in the 8K and HSS fractions was similar to that in the corresponding fractions of visual cortex, but the RNA in b-ribosomes was considerably greater and that in f-ribosomes considerably less than in the brain tissue. This distribution of RNA between the b- and f-ribosomes is in conformity with observations that a majority of liver ribosomes are found attached to the membrane (Andrews and Tata, 1971) whereas only 10–15% of the total

Table I. Distribution of RNA in the Subcellular Fractions of the Visual Cortex and Liver of Dark-Reared and Light-Exposed Rats[a]

Treatment	RNA/DNA in homogenate as % of N4	% of total RNA recovered in fractions			
		8K	b-r	f-r	HSS
Neuron					
N4	100	24.5	12.2	51.2	12.1
D3	65	17.5	5.1	38.2	39.2
D4	52	15.2	4.3	35.4	45.1
D3-L1	72	18.5	5.7	47.1	28.7
Glia					
N4	100	29.8	10.4	48.3	11.5
D3	71	24.2	9.1	38.1	28.6
D4	57	22.8	8.9	34.2	34.1
D3-L1	75	27.2	9.6	46.5	16.7
Liver					
N4	100	30.1	33.2	26.2	10.5
D3	94	29.8	34.5	25.8	9.9
D4	95	31.2	31.2	27.1	10.5
D3-L1	102	28.9	33.1	26.6	11.4

[a] See Section B, Materials and Methods and Fig. 1 for explanation of symbols and preparation of tissue fractions.

ribosomal population in brain are bound to endoplasmic reticulum (Merits et al., 1969; Murthy, 1970, 1972a). The high proportion of RNA found in the 8K fraction is presumably due to sedimentation of b-ribosomes, since purified nuclei, mitochondria, and synaptosomes contribute only a small fraction of total cellular RNA. It has, in fact, been reported for several tissues that an appreciable fraction of b-ribosomes are pelleted under conditions used to separate nuclei, mitochondria, and, in the case of brain, synaptosomes (Murthy, 1970; Adelman et al., 1973; O'Toole and Pollack, 1974; Lewis and Tata, 1973).

In addition to the decrease in the amount of RNA per cell (Table I, RNA/DNA ratio), the relative proportion of RNA in the different subcellular fractions of the visual cortex also appeared to undergo pronounced changes as a result of dark rearing. Following 3 weeks of this treatment, the RNA in the HSS fraction increased 3-fold in neurons and 2.5-fold in glial cells. This increase occurred at the expense of RNA in the particulate fractions. In the case of neurons, all three fractions, 8K, b-R, and f-R lost their RNA; in glia, the 8K and b-R fractions were not appreciably affected, with the f-R fraction being the major site of RNA loss. Extension of the dark period from 3 to 4 weeks led to a slightly further aggravation of the situation in the same direction, i.e., accumulation of RNA in the soluble fraction at the expense of the particulate fractions. Exposure of the D3 rats to 1 week of light/dark cycle (D3-L1) increased the RNA content of the f-ribosomes in glia, thus bringing the RNA distribution in all fractions of glia very close to the normal pattern. In neurons also, light exposure resulted in an increase in f-ribosomal RNA, but the membrane-bound RNA in either the b-R or 8K fractions remained quite low, indicating that, although ribosomes were being produced by new synthesis or by reconstitution of previously existing precursors, these ribosomes were unable to attach themselves to the neuronal endoplasmic reticulum. The lack of capacity for membrane–ribosome interaction in the neurons could be due to a structural deficiency in the attachment sites or related areas either in the ribosomes or in the membrane or both.

In common with the total RNA content, the relative RNA distribution in the subcellular fractions of liver remained unaffected by dark rearing or subsequent exposure of the animals to light/dark cycle.

In order to investigate whether and to what extent the production of ribosomes and their efficiency to synthesize proteins are affected by dark rearing and subsequent light exposure, the rats were injected intraperitoneally with a mixture of 50 uCi each of ^{3}H-phenylalanine and ^{14}C-uridine dissolved in physiological saline. After 4 hr, the rats were sacrificed and radioactivity was determined in various subcellular fractions of the visual cortex and liver (Fig. 1). A small portion of the homogenate was precipitated with 5% TCA in order to determine acid soluble radioactivity. After centrifugation, TCA was removed from the supernatant by repeated extraction with ether. The b-R, f-R, and HSS fractions were also similarly precipitated with TCA. The sediments were dissolved in NCS solubilizer for counting. The results show that the amount of ^{3}H radioactivity in the acid-soluble pool of either neurons or glia was affected only slightly by 3 or 4 weeks of darkness (Table II), indicating that the characteristics of the blood–brain barrier, at least for phenylalanine, were intact and that the transport of this amino acid into neurons and glia of the visual cortex was near normal. In the normal animal (N4), about 16% of the radioactivity of this pool was incorporated into soluble proteins (HSS) and 0.82 and 3.9% into b- and f-ribosomes, respectively. The apparently high incorporation of radioactivity in the HSS is due to the fact that this fraction contains the highest proportion of cytoplasmic protein (Murthy and Rappoport, 1965). Following 3 or 4 weeks of dark rearing, the percent of acid-soluble radioactivity incorporated into b- and f-ribosomes were diminished by 50 and 35%, respectively, in neurons. The percent of radioactivity incorporated into

Table II. Incorporation in Vivo of ^{3}H-phenylalanine into RNP of Ribosomal and HSS Fractions of Visual Cortex and Liver of Dark-Reared and Light-Exposed Rats

Treatment	DPM in acid-soluble pool as % of N4	DPM in RNP as % of pool DPM			Relative specific activity (RSA)[a] as % of N4		
		b-r	f-r	HSS	b-r	f-r	HSS
Neuron							
N4	100	0.82	3.9	15.9	100	100	100
D3	82	0.42	2.8	18.0	82	79	58
D4	85	0.41	2.6	17.2	84	75	55
D3-L1	105	0.52	3.2	16.8	90	84	75
Glia							
N4	100	1.2	5.1	18.2	100	100	100
D3	89	1.1	3.5	19.7	79	81	65
D4	92	0.92	3.4	19.1	75	82	62
D3-L1	109	1.1	4.9	18.6	85	85	78
Liver							
N4	100	3.5	4.2	24.5	100	100	100
D3	98	3.2	4.1	25.2	92	105	95
D4	105	3.3	4.3	26.4	95	97	106
D3-L1	100	3.6	4.0	25.1	102	109	92

[a] Relative specific activity represents the disintegrations per minute per milligram (dpm) ribonucleoprotein (RNP) corrected for the concentration of radioactivity in the acid soluble fraction of a given tissue. It is given by the equation: RSA = Dpm in RNP of a given sample of tissue/DPM in the acid-soluble pool of the same sample.

the HSS fraction was, however, unaffected. This would indicate that the pool radioactivity was not being used as efficiently as in the control animal for the formation of complete ribosomes, especially b-ribosomes in the D3 and D4 rats. The radioactivity in the HSS fraction, under these conditions, could represent the newly formed polypeptides as well as ribosomal precursor proteins which were not stabilized or integrated into ribosomal particles. When rats reared in the dark for 3 weeks were exposed to light/dark cycle for one week (D3-L1) there was an improvement in the utilization of pool phenylalanine for ribosome synthesis, but this was less marked for the b-ribosomes than for the f-ribosomes.

[3]H-Radioactivity incorporated into glial ribosomes (both f- and b-ribosomes) and HSS of normal optical cortex (N4) was significantly higher than for the corresponding neuronal fractions, when measured as percent of acid-soluble radioactivity (Table II). In addition, dark rearing of rats for 3 or 4 weeks had no significant effect on the level of incorporation in the b-ribosomes. Only the f-ribosomes suffered a decline in incorporation which, however, was restored almost to the

normal level by one week of exposure to light/dark cycle (D3-L1).

When incorporations into various subcellular fractions were calculated as relative specific activity (RSA), i.e., DPM/mg RNP, after correction for pool radioactivity, it was observed that the RSAs of b- and f-ribosomes of either neurons or glia were affected much less by dark rearing than the soluble fractions of these two types of cells (Table II). The RSAs of the HSS fractions of D3 and D4 rats were decreased by about 40% in both neurons and glia. However, this decrease can be accounted for by the fact that the concentration of RNA, and presumably of ribonucleoprotein, in the HSS of the optical cortex of these animals increased by dark treatment at the expense of the ribosomes as shown in Table I. Since calculation of RSA was based on unit weight of RNP, this would be expected to result in decreased specific activity for the soluble fraction and slightly enhanced values for the ribosomes.

In contrast to brain cells, liver was unaffected by light deprivation or subsequent exposure to light. Neither the acid-soluble pool, nor the percent of pool radioactivity incorporated into the subcellular fractions nor the relative specific activities of these fractions were changed in any way by these treatments.

In the above experiment, ^{3}H-phenylalanine and ^{14}C-uridine were injected simultaneously in order to study the synthesis of cytoplasmic ribonucleoprotein complexes as well as soluble proteins in rats subjected to different treatments. Incorporation of uridine into the various subcellular fractions of optical cortex differed from the incorporation of phenylalanine in certain parameters and was similar in others (Table III). Thus, in the normal rat (N4), while the highest incorporation of phenylalanine (as percent of acid-soluble radioactivity) occurred in the HSS fraction and much less in the ribosomes, the highest incorporation of uridine occurred in the ribosomes and much less in the HSS fraction. As in the case of phenylalanine incorporation, uridine incorporation was also 5–6 times higher in f-ribosomes than in b-ribosomes. These results are predictable since the highest proportion of cytoplasmic protein in brain is found in the soluble fraction, whereas the highest proportion of cytoplasmic RNA is present in the ribosomes (Murthy and Rappoport, 1965). The greater incorporation of both phenylalanine and uridine into f-ribosomes as compared to b-ribosomes is also in conformity with the observation that b-ribosomes form only 15–20% of the total ribosomal population of the brain cells. In liver, incorporations of phenylalanine and uridine into b- and f-ribosomes were not very different, in parallel with the findings that the b-ribosomes constitute a majority of the total ribosomes of this tissue.

Table III. Incorporation in Vivo of ^{14}C-Uridine into RNP of Ribosomal and HSS Fractions of Visual Cortex and Liver of Dark-Reared and Light-Exposed Rats[a]

Treatment	DPM in acid-soluble pool as % of N4	DPM in RNP as % of pool DPM			Relative specific activity (RSA) as % of N4		
		b-r	f-r	HSS	b-r	f-r	HSS
Neuron							
N4	100	3.0	18.5	2.3	100	100	100
D3	86	1.5	13.3	4.2	85	75	61
D4	83	1.2	12.5	5.1	87	78	52
D3-L1	86	1.6	16.8	3.5	97	86	75
Glia							
N4	100	4.5	20.5	3.1	100	100	100
D3	90	4.0	18.2	2.9	80	74	64
D4	85	3.6	14.8	3.0	78	75	58
D3-L1	87	4.4	19.2	3.0	86	80	80
Liver							
N4	100	7.1	10.5	6.5	100	100	100
D3	96	7.0	10.8	6.8	95	105	95
D4	99	7.3	9.3	6.2	98	107	101
D3-L1	104	7.9	11.1	7.0	105	98	97

[a] The procedures were the same as in Table II and the results were obtained from the same series of experiments using simultaneous labeling with ^{3}H-phenylalanine and ^{14}C-uridine. The above values were calculated from ^{14}C counts.

As in the case of phenylalanine, incorporation of uridine into neurons was affected by dark rearing more drastically at the level of b-ribosomes than at the level of f-ribosomes, although both were considerably diminished. Exposure to 1 week of light/dark cycle following 3 weeks of darkness (D3-L1), led to an almost complete recovery of f-ribosome synthesis, whereas the synthesis of b-ribosomes was only partially restored. In contrast to the situation existing at the ribosomal level, incorporation of uridine into the acid-insoluble fraction of HSS increased by about 2-fold when rats were reared under darkness for 3 to 4 weeks. This increase in uridine radioactivity cannot be due to an increased synthesis of tRNA since there is no reason to believe that transcription of tRNA genes would be stimulated while at the same time ribosomal RNA genes are suppressed. It is more probable that this increased radioactivity is due to ribosomal precursor RNP particles and subunits which are unable to form intact ribosomes and cannot bind to the endoplasmic reticular membrane and consequently remain as part of the nonsedimentable fraction.

In contrast to neurons, synthesis of f-ribosomes in the glial cells of optical cortex was only slightly diminished and those of b-ribosomes and of cytoplasmic RNP were not changed at all after 3 or 4 weeks of darkness (D3 and D4). Even the synthesis of glial f-ribosomes returned to normal after exposure of the dark reared rats to 1 week of light/dark cycle (D3-L1). Incorporation of uridine into the subfractions of liver was not altered by any of the above treatments.

The relative specific activities of b-R, f-R, and HSS fractions, based on ^{14}C-uridine, were very similar to those based on ^{3}H-phenylalanine. The decrease in ^{14}C-RSA of HSS ribonucleoproteins, again, was in conformity with the previous observation (Table I) that acid-insoluble material (presumably proteins, RNA, and ribonucleoprotein complexes) accumulated in the soluble fraction at the expense of the particulate ribosomes and hence contributed to decreased RSA values.

The lack of effect of sensory deprivation or overstimulation on a nonneural tissue such as liver has been observed by several investigators. However, within the brain itself, imposition of a given modality of sensory input has been reported to affect not only the cortical region directly involved with its reception and processing, but also certain other areas of the cortex. For example, according to Rose (1967), when rats were exposed to darkness for 50 days after birth and then exposed to light, there was an enhanced incorporation of ^{3}H-lysine into proteins of both visual and frontal cortex, but not into liver proteins. Appel et al. (1967) found that 15 min of light after 3 days of darkness produced an increased number of polysomes and polysomal RNA in both visual and remaining cortex. In mice, reared for prolonged periods in darkness, the auditory cortex was found to show hypertrophy, even after normalization had occurred in the visual cortex (Gyllensten et al., 1966). This phenomenon has been explained in terms of reduction of associative stimuli, hormonal effects mediated through the hypothalamus, and increased reliance on a complementary region of the cerebral cortex under conditions when the primary region atrophies due to lack of sufficient sensory input. We, therefore, examined the relative effects of light deprivation and subsequent stimulation on the synthesis of RNA and protein in the visual cortex as compared to the rest of the cerebral cortex of rat (V/C ratio) (see Table IV).

In the normal rat (N4), incorporations of ^{14}C-uridine and ^{3}H-phenylalanine into ribosomes and soluble RNP of visual cortex were not much different from those in the rest of the cortical area of brain, whether the subcellular fractions were derived from the neurons or from

Table IV. Incorporation in Vivo of ^{3}H-Phenylalanine and ^{14}C-Uridine into RNP of Ribosomal and HSS Fractions of Visual Cortex as Compared to Other Cortical Regions of Dark-Reared and Light-Exposed Rats

| | V/C ratio[a] | | | | | |
| | b-Ribosomes | | f-Ribosomes | | HSS | |
Treatment	Phe	Uridine	Phe	Uridine	Phe	Uridine
Neuron						
N4	0.98	1.05	1.12	1.08	1.10	1.15
D3	0.51	0.47	0.67	0.62	0.78	0.82
D4	0.42	0.40	0.61	0.58	0.81	0.87
D3-L1	0.61	0.57	0.78	0.81	0.78	0.75
Glia						
N4	1.20	1.16	1.06	1.11	1.05	1.07
D3	0.88	0.85	0.63	0.71	0.88	0.87
D4	0.81	0.77	0.58	0.61	0.85	0.81
D3-L1	0.92	0.89	0.89	0.85	0.91	0.92

[a] For calculation of V/C ratios, the relative specific activities (RSA) for each of the fractions were used. See Table II for definition of RSA and description of other experimental procedures.

glia. The ratios of incorporation into the visual cortex and that in the rest of the cerebral cortex (V/C ratio) were approximately equal to one. But in the D3 and D4 rats, incorporation of these precursors fell by 50% in the b-R, f-R, or HSS fractions of the neurons of the visual cortex as compared to similar fractions from the rest of the cerebral cortex. The glial cells of the visual cortex also showed a decline in the V/C ratios, but this decline was not as pronounced as in neurons.

In view of previous observations that the proportion of total cellular RNA as well as newly synthesized RNA accumulated in the soluble fraction at the expense of ribosomes (Tables I and III), it was of interest to determine the capacity of b- and f-ribosomes still present intact in the brain cells to carry out protein synthesis. For this purpose, we measured the incorporation of phenylalanine by various ribosomal preparations of the visual cortex in the presence and in the absence of added poly-U. The ratio of synthesis of polyphenylalanine (with poly-U) to that of protein (with endogenous mRNA) (poly-P/protein ratio) was taken as an indirect measure of the monomer to polysomal ratio in a given population of ribosomes. It is seen from Table V that the poly-P/protein ratios of neuronal b-ribosomes and f-ribosomes increased nearly 4-fold and 2-fold, respectively, after 3–4 weeks of dark rearing,

Table V. Incorporation in Vitro of ^{3}H-Phenylalanine into Protein and
poly-P by f- and b-Ribosomes of Visual Cortex
of Dark-Reared and Light-Exposed Rats[a]

	Incorporation			
	b-Ribosomes		f-Ribosomes	
Treatment	Poly-P/ protein ratio	Increase over N4	Poly-P/ protein ratio	Increase over N4
Neuron				
N4	0.53	1.0	0.85	1.0
D3	1.82	3.4	1.52	1.8
D4	2.10	4.0	1.75	2.0
D3-L1	1.85	3.5	1.18	1.4
Glia				
N4	0.41	1.0	0.72	1.0
D3	0.75	1.8	1.68	2.3
D4	1.10	2.7	2.10	3.0
D3-L1	0.58	1.4	1.40	2.0

[a] Reaction conditions for amino acid incorporation by brain ribosomes and calculation of poly-P/protein ratios were as described previously (Murthy, 1966). Radioactivity due to poly-P synthesis was calculated by subtracting the disintegrations per minute incorporated per milligram ribosomal protein in the absence of added poly-U (protein) from that obtained in the presence of poly-U (protein + poly-P).

indicating that the relative proportion of monomers, inactive in endogenous protein synthesis, increased as a result of this treatment. Glial b- and f-ribosomes also showed similar increases in this ratio, but in contrast to neurons, the f-ribosomes of glia responded more severely to darkness than b-ribosomes. Exposure of the dark-reared animals to 1 week of light/dark cycle (D3-L1) was partially effective in restoring the normal proportion of f-ribosomes of neurons and b-ribosomes of glia. The b-ribosomes of neurons and f-ribosomes of glia did not recover their normal values by this treatment.

The decrease in the proportion of neuronal and glial polysomes (Table V) and a simultaneous diminution of incorporation of ^{14}C-uridine and ^{3}H-phenylalanine into ribosomes (Tables II and III) indicated that transcription of both ribosomal RNA and mRNA were probably inhibited by subjecting the newborn rats to prolonged darkness. In order to verify the relative effect of light deprivation on the synthesis of these two types of RNA, newly synthesized total polysomal RNA, labeled *in vivo* with ^{14}C-uridine, was separated into two fractions, one enriched in ribosomal RNA (pH 6 RNA) and the other enriched in cytoplasmic

Table VI. Incorporation in Vivo of ^{14}C-Uridine into pH 6 RNA and
pH 8 RNA of f- and b-Ribosomes from the
Visual Cortex of Dark-Reared and Light-Exposed Rats[a]

| | (DPM/mg) × 10^{-3} | | | | | | | |
| | b-Ribosomes | | | | f-Ribosomes | | | |
Treatment	pH 6 RNA	%	pH 8 RNA	%	pH 6 RNA	%	pH 8 RNA	%
Neuron								
N4	13.1	100	25.6	100	12.2	100	18.5	100
D3	8.1	62	13.1	51	8.5	70	9.3	50
D4	6.6	50	8.2	32	7.9	65	7.2	39
D3-L1	8.9	68	21.8	85	9.8	80	16.3	88
Glia								
N4	14.8	100	26.2	100	11.8	100	16.4	100
D3	12.4	84	16.0	61	8.5	72	9.8	60
D4	10.7	72	14.4	55	8.0	68	8.5	52
D3-L1	13.0	88	24.1	92	10.6	90	15.4	94

[a]pH 6 and pH 8 RNAs were isolated from b- and f-ribosomes as described in Section B,
Materials and Methods. The incorporated radioactivity of a given RNA was corrected
for acid soluble radioactivity.

mRNA (pH 8 RNA). Table VI indicates that the synthesis of both
pH 6 and pH 8 RNAs were inhibited by rearing rats in darkness for
3 to 4 weeks. This was true for both b- and f-ribosomes of both neuronal
and glial cells. However, the following differences were noted: (a) the
pH 8 RNA synthesis suffered a more drastic inhibition than the pH 6
RNA, for example, nearly 70% for pH 8 RNA as compared to 50% for
pH 6 RNA of neuronal b-ribosomes, after 4 weeks of darkness; (b) when
the dark-reared animals were brought into the normal light/darkness
cycle for 1 week, there was only a partial recovery of pH 6 RNA synthesis whereas mRNA synthesis was almost completely restored to the
normal value; (c) the above phenomena were very similar in the b- and
f-ribosomes, and (d) the effects of dark rearing were less pronounced
and the recovery following light exposure was more complete for the
polysomal RNAs of glia as compared to those of neurons.

Published reports from various laboratories indicate that the degree
and nature of response of dark-reared animals to reintroduction of light,
as judged by various biochemical reactions implicated in protein and
nucleic acid synthesis, depend on such factors as the age of the animal,
the duration of darkness, as well as the qualitative and quantitative

parameters of light exposure. For example, in their studies on the bio-chemical correlates of imprinting in chicks, Bateson et al. (1973) found that the incorporation of uracil into RNA of chick brains during exposure to flashing light was most rapid during the first few hours and then decreased with prolonged exposure. Singh and Talwar (1967, 1969) have reported that intermediate levels of light intensity increased protein synthesis in the cortex while extreme levels tended to depress it. Further, these authors as well as others (Dewar and Reading, 1970; Bateson and Wainwright, 1972; Simner, 1973, 1974), have demonstrated that flicker-ing or flashing light was more effective than continuous light in inducing enhanced neurochemical activity and in eliciting specific behavioral patterns in a number of different species of animals. Macromolecular synthesis as well as expression of innate behavior also appear to be influenced by the color of light to which the animal is exposed (Hess, 1956; Hess and Gogel, 1954; Singh and Talwar, 1967). Many of these experiments were performed using normal animals that were not de-prived of light stimulation over a prolonged period. According to results presented in this paper (Tables II, III, and VI), return of rats previously subjected to 3 weeks of darkness to a normal light/dark cycle for 1 week enabled a partial recovery in terms of enhanced polysomal protein and RNA synthesis in neurons and glia. In view of the literature cited above, we investigated the effects of brief periods of light exposure on polysomal RNA and protein synthesis in the visual cortex of rats reared in darkness from the time of birth. Table VII shows that exposure to continuous white, blue, or red light for 4 hr did not lead to an apprecia-ble enhancement of phenylalanine incorporation into ribosomal pro-tein or uridine incorporation into ribosomal RNA, irrespective of the nature of the ribosomes (i.e., f- or b-ribosomes) or the type of the cell examined (neurons or glia). Only the synthesis of neuronal pH 8 RNA was increased by continuous white light, this effect being more pro-nounced for f-polysomes than for b-polysomes. Flickering lights of all colors were superior to continuous light in this respect, but again, the white flickering light gave a much higher response than either the blue or red lights. The highest incorporation in response to flickering light occurred in the pH 8 RNA and to a lesser extent in the proteins and in the pH 6 RNA. f-Ribosomes were more affected than b-ribosomes and neurons more than glia.

Since light exposure of dark-reared animals had the most pro-nounced effect on the synthesis of messenger-enriched fractions of polysomal RNAs, it was of interest to see if there exist any differences in the base composition of mRNAs in the visual cortex of normal and experimental rats. Since the pH 8 RNA used in the earlier experiments,

Table VII. Effect of Light Stimuli of Different Wavelengths on
[3]H-Phenylalanine and ^{14}C-Uridine Incorporation into
Ribosomal Proteins and RNAs of Visual Cortex
of Rats Previously Deprived of Light

| | RSA as % of D4 | | | | | |
| | b-Ribosomes | | | f-Ribosomes | | |
Treatment[a]	Protein	pH 6 RNA	pH 8 RNA	Protein	pH 6 RNA	pH 8 RNA
Neuron						
D4 + W_c	105	116	136	115	112	152
D4 + W_f	185	156	258	250	144	340
D4 + B_c	100	108	105	118	108	132
D4 + B_f	162	138	185	185	131	250
D4 + R_c	100	108	116	108	102	112
D4 + R_f	144	133	162	149	118	189
Glia						
D4 + W_c	102	99	115	110	109	113
D4 + W_f	142	126	186	172	120	258
D4 + B_c	98	104	110	105	104	110
D4 + B_f	130	121	162	143	120	235
D4 + R_c	100	93	98	105	99	110
D4 + R_f	115	105	138	122	115	185

[a] Rats were exposed to continuous (subscript c) or flickering (subscript f) white (W),
blue (B) or red (R) lights for periods of 4 hr using the procedure described in Section B,
Materials and Methods. Immediately before the start of light stimulation, each animal
was injected intraperitonially with a mixture of 50 μCi of each of L-^{3}H-phenylalanine
and ^{14}C-uridine in 0.5 ml of physiological saline. At the end of 4 hr, the rats were
killed, the various RNA and protein fractions were prepared and radioactivities
determined as detailed in Section B, Materials and Methods.

although enriched in mRNA, still contained considerable amounts of
ribosomal RNA (Campagnoni et al., 1971), it was purified by affinity
chromatography on oligo-dT cellulose (see Section B, Materials and
Methods). Results presented in Table VIII show that there were considerable differences in base composition of mRNAs depending upon
the nature of the polysomes, the cell type from which they were derived,
and the treatment undergone by the animals. Zomzely et al. (1970) have
reported U/C and A/G ratios of 1.59 and 0.85, respectively, for polysomal mRNA. This corresponds well with the composition of the free
polysomal mRNA of normal rat visual cortex, as found in our experiments. However, even in the normal rat visual cortex, b-polysomal
mRNA of neurons as well as the f- and b-polysomal mRNAs of glial

Table VIII. Base Composition of mRNAs of f- and b-Ribosomes
from the Visual Cortex of Dark-Reared and Light-Exposed Rats

| | Percentage base composition [a] | | | |
| | b-Ribosomes | | f-Ribosomes | |
Treatment	A/G	U/C	A/G	U/C
Neuron				
N4	0.82	1.31	0.75	1.51
D4	0.85	1.48	0.78	1.32
D4-WF	0.80	1.61	0.85	1.65
Glia				
N4	0.84	1.20	0.78	1.35
D4	0.75	1.38	0.72	1.42
D4-WF	0.93	1.52	0.86	1.55

[a] Base composition of mRNA was determined by tritium postlabeling, as described in Section B, Materials and Methods.

cells deviated significantly from this value. Thus, in general, glial mRNAs exhibited lower U/C and higher A/G ratios than neuronal mRNAs. Similarly b-polysomal mRNAs of either glia or neurons had lower U/C and higher A/G ratios than f-polysomal mRNAs. Four weeks of dark rearing or 1 week of subsequent exposure to light also produced changes in these values, the natures of which were dependent on the type of cells and of polysomes involved. In all cases, exposure of rats previously reared in darkness to flickering white light produced mRNA with U/C ratios significantly higher than those of either the control animals or dark-reared animals. These differences in base composition can be assumed to reflect an alteration in the rates and types of proteins synthesized under the modulating influence of dark and light treatments. In fact, that such altered synthesis of protein takes place in response to light has been demonstrated by Richardson and Rose (1973a) for retinal proteins. These authors measured lysine incorporation in 41 fractions and found significant differences in only 13 of these fractions after 1 hr of exposure. In another study, they also reported differential protein synthesis in the visual cortex (Richardson and Rose, 1973b). Singh and Talwar (1969) have also demonstrated that the brain-specific protein S-100 was among the group of low molecular weight proteins rapidly and selectively synthesized in the visual cortex in response to flickering light.

D. CONCLUSIONS

The results presented in this paper show that, when newborn rats are deprived completely of light stimulation during the first 3–4 weeks immediately after birth, the following biochemical changes occurred in neuronal and glial cells of the visual cortex: (a) the RNA content per cell was decreased; (b) the relative distribution of RNA in the subcellular fractions of the cell underwent drastic changes in such a manner that RNA accumulated in the soluble fraction at the expense of particulate, ribosomal RNA; (c) although entry of injected radioactive phenylalanine or uridine into the acid-soluble pool was only slightly affected, the percentages of pool radioactivity incorporated into the membrane-bound and free ribosomes diminished for either of these two precursors while the percentages incorporated into the soluble fraction increased; (d) the ability of the polysomes to synthesize protein *in vitro* using endogenous RNA was reduced, with a simultaneous accumulation of monomeric ribosomes; (e) the incorporation *in vivo* of uridine into both ribosomal and mRNA decreased and (f) the base composition of mRNA associated with polysomes changed. The extent of these changes depended upon such factors as the duration of darkness, the type of cell, and the subcellular fraction under consideration. As examples, 4 weeks of darkness generally led to more pronounced effects than 3 weeks, and neurons were more severely affected than the glia. In neurons, the membrane-bound ribosomes were more susceptible to deprivation of light than free-ribosomes whereas the reverse was the case in glia.

When rats reared up to 3 weeks in darkness were returned to light/dark cycle for 1 week, the normal patterns of the above biochemical parameters were only partially restored, the extent of reversibility again depending upon the cell type and the subcellular fraction. Neurons were less able to recover than glia. Membrane-bound polysomes of neurons were more refractory to repair by this treatment than free polysomes.

The degeneratory biochemical changes noted above appeared to be mostly confined to the visual cortex. Other cerebral cortical regions or an extracerebral tissue such as liver were unaffected.

The greater susceptibility of neurons to light deprivation as compared to glia and their lower capacity for recovery when the light stimulation was resumed suggested that these differences are probably related to their different roles in brain function. Since neurons are primarily concerned in the establishment of coordinated interconnecting networks in response to environmental stimuli by orderly elongations of axonal and dendritic extensions, it is reasonable to assume that their morphology and biochemistry are more closely dependent on sensory input than those of glia. The observation that, in neurons, the membrane-bound ribosomes are more severely affected than free ribosomes indicated that the former could have a special role in the utilization or processing of sensory information. Since membrane-bound ribosomes are believed to take part in the synthesis of secretory proteins, one of their possible roles during neuronal activity could be the elaboration of proteins that are transferred from the soma to the periphery by axoplasmic transport. Our data also show that during the recovery period following resumption of light stimulus, the free ribosomes increased with no parallel increase in membrane-bound ribosomes. One explanation for this could be that light deprivation damaged the membrane more severely than the biochemical machinery involved in ribosome formation, with the result that the ribosomes were prevented from attachment to membrane. In support of this possibility, some preliminary data obtained in our laboratory indicate that the incorporation of fatty acids into brain endoplasmic reticular membrane is also inhibited by light deprivation.

E.   SUMMARY

Rats were maintained in darkness from birth up to 3 or 4 weeks of age and were then exposed to light for 1 week. Neuronal- and glial-enriched cells were separated from homogenates of the visual cortex. Free and membrane-bound polysomes were isolated from each cell type. The relative concentrations of free and membrane-bound polysomes and the capacity of incorporation *in vivo* of ^{3}H-phenylalanine and ^{14}C-uridine into proteins and RNA of different subcellular fractions were measured. Amino acid incorporation *in vitro* by free and membrane-bound polysomes was also determined in the presence and in the absence of poly-U. Some of these experiments were carried out with brain regions other than the visual cortex and in liver for comparative purposes. The results indicate that light deprivation decreased the level of cytoplasmic RNA. The distribution of RNA between the various subcellular fractions also changed in such a manner that a larger proportion of RNA became nonsedimentable in the light-deprived rats as compared to normal animals. Membrane-bound polysomes suffered the greater damage, both in RNA and in the capacity for protein synthesis. It appeared that this resulted not only from the degradation of polysomal components, but also possibly of the membrane structure. There was a partial recovery when the rats were exposed to light for 1 week, but the values remained lower than those of normal animals.

In another series of experiments, light-deprived rats were exposed to brief periods of continuous or intermittent blue, red, or white light. Simultaneous administration of radioactive phenylalanine and uridine showed that intermittent light led to a higher incorporation of these precursors than continuous light. The white light was superior in this respect to the colored lights. Neurons reacted more severely than glial cells to both light deprivation and light exposure. The base composition of mRNA, isolated by oligo-dT cellulose chromatography, was also found to differ among polysomes from normal, light-deprived, and light-exposed rats.

F. REFERENCES

Adelman, M. R., Blobel, G., and Sabatini, D. D. (1973). An improved cell fractionation procedure for the preparation of rat liver membrane bound ribosomes. *J. Cell. Biol.* 56, 191–205.

Altman, J., and Das., G. D. (1964). Autoradiographic examination of the effects of enriched environment on the rate of glial multiplication in the adult rat brain. *Nature (London)* 204, 1161–1163.

Andrews, T. M., and Tata, J. R. (1971). Protein synthesis by membrane-bound and free ribosomes of secretory and nonsecretory tissues. *Biochem. J.* 121, 683–694.

Appel, S. H., Davis, W., and Scott, S. (1964). Brain polysomes: Response to environmental stimulation. *Science* 157, 836–838.

Aviv, H., and Leder, P. (1972). Purification of biologically active globin messenger RNA by chromatography on oligothymidylic acid-cellulose. *Proc. Nat. Acad. Sci., U.S.* 69, 1408–1412.

Bateson, P. P. G., Rose, S. P. R., and Horn, G. (1973). Imprinting: Lasting effects on uracil incorporation into chick brain. *Science* 181, 576–578.

Bateson, P. P. G., and Wainwright, A. A. P. (1972). The effects of prior exposure to light on the imprinting process in domestic chicks. *Behaviour* 42, 279–290.

Bloemendal, H., Bout, W. S., DeVries, M., and Benedetti, E. L. (1967). Isolation and properties of polyribosomes and fragments of the endoplasmic reticulum from rat liver. *Biochem. J.* 103, 177–182.

Brattgard, S. O. (1952). The importance of adequate stimulation or the chemical composition of retinal ganglion cells during early post-natal development. *Acta Radiol. Suppl.* 96, 1–80.

Burke, W., and Hayhow, W. R. (1968). Disuse in the lateral geniculate nucleus of the cat. *J. Physiol. (London)* 194, 495–519.

Burton, K. (1956). A study of the conditions and mechanism of the diphenylamine reaction for the colorimetric estimation of DNA. *Biochem. J.* 62, 315–323.

Campagnoni, A. T., Dutton, G. R., Mahler, H. R., and Moor, W. J. (1971). Fractionation of the RNA components of rat brain polysomes. *J. Neurochem.* 18, 601–611.

Chow, K. L., and Stewart, D. L. (1972). Reversal of structural and functional effects of long term visual deprivation in cats. *Exp. Neurol.* 34, 409–433.

Darnell, J. E., Philipson, L., Wall, R., and Adesnik, M. (1971). Polyadenylic acid sequences: Role in conversion of nuclear RNA into messenger RNA. *Science* 174, 507–510.

Dewar, A. J., and Reading, H. W. (1970). Nervous activity and RNA metabolism in the visual cortex of rat brain. *Nature (London)* 225, 869–870.

Diamond, M. C., Law, F., Rhodes, H., Lindner, B., Rosenzweig, M. R., Krech, D., and Bennett, E. L. (1966). Increases in cortical depth and glia numbers in rats subjected to enriched environment. *J. Comp. Neurol.* 128, 117–125.

Fleck, A., and Begg, D. (1965). The estimation of ribonucleic acid using ultraviolet absorption measurements. *Biochim. Biophys. Acta* 108, 333–339.

Globus, A. (1975). In: *The Developmental Neuropsychology of Sensory Deprivation* (A. H. Riesen, ed.), pp. 9–91, Academic Press, New York.

Gorski, J., Morrison, M. R., Merkel, C. G., and Lingrel, J. B. (1974). Size heterogeneity of polyadenylate sequences in mouse-globin messenger RNA. *J. Mol. Biol.* 86, 363–371.

Gyllensten, L., Malmfors, R., and Norrlin, M. L. (1966). Growth alteration in the auditory cortex of visually deprived mice. *J. Comp. Neurol.* 126, 463–469.

Hadjivassiliou, A., and Brawerman, G. (1967). Template and ribosomal ribonucleic acid components in the nucleus and the cytoplasm of rat liver. *Biochemistry* 6, 1934–1940.

Hallinan, T., Murty, C. N., and Grant, J. H. (1968). Early labeling with glucosamine-^{14}C of granular and agranular endoplasmic reticulum and free ribosomes from rat liver. *Arch. Biochem. Biophys.*, 125, 715–720.

Hess, E. H., and Gogel, W. C. (1954). Natural preferences of the chick for objects of different colors. *J. Psychol.* 38, 483–493.

Hess, E. H. (1956). Natural preferences of chicks and ducklings for objects of different colors. *Psychol. Rep.* 2, 477–483.

Higashi, T., Kudo, H., and Kashiwagi, K. (1972). Specific precipitation of catalase-synthesizing ribosomes by anticatalase antiserum. *J. Biochem. (Tokyo)* 71, 463–470.

Hubel, D. H., and Wiesel, T. N. (1970). The period of susceptibility to the physiological effects of unilateral eye closure in kittens. *J. Physiol.* (*London*) 206, 419–436.

Lewis, J. A., and Tata, J. R. (1973). A rapidly sedimenting fraction of rat liver endoplasmic reticulum. *J. Cell. Sci.* 13, 447–459.

Lowry, O. H., Rosenbrough, N. J., Farr. A. L., and Randall, R. J. (1951). Protein measurement with the Folin phenol reagent. *J. Biol. Chem.* 193, 265–275.

Merits, I., Cain, J. C., Razok, E. J., and Minard, F. N. (1969). Distribution between free and membrane-bound ribosomes in rat brain. *Experientia* 25, 739–740.

Morrison, M. R., and Lingrel, J. B. (1975). Characterization of globin messenger ribonucleic acids in membrane polysomes of mouse reticulocytes. *J. Biol. Chem.* 250, 848–852.

Murthy, M. R. V., and Rappoport, D. A. (1965). Biochemistry of the developing rat brain. V. Cell-free incorporation of L-[1-^{14}C]leucine into microsomal protein. *Biochim. Biophys. Acta* 95, 121–131.

Murthy, M. R. V. (1966). Protein synthesis in growing rat tissues. 1. Effect of various metabolites and inhibitors on phenylalanine incorporation by brain and liver ribosomes. *Biochim. Biophys. Acta* 119, 586–598.

Murthy, M. R. V. (1970). Membrane-bound and free ribosomes in the developing rat brain. In: *Protein Metabolism of Nervous System* (A. Lajtha, ed.), pp. 109–127, Plenum Press, New York.

Murthy, M. R. V. (1972a). Free and membrane-bound ribosomes of rat cerebral cortex: Protein synthesis *in vivo* and *in vitro*. *J. Biol. Chem.* 247, 1936–1943.

Murthy, M. R. V. (1972b). Free and membrane-bound ribosomes of rat cerebral cortex: Metabolism of ribosomal and messenger ribonucleic acid. *J. Biol. Chem.* 247, 1944–1955.

Nihei, T. (1971). *In vitro* amino acid incorporation into myosin by free polysomes of rat skeletal muscle. *Biochem. Biophys. Res. Commun.* 43, 1139–1149.

O'Toole, K., and Pollack, J. K. (1974). Changes in free and membrane-bound ribosomes during development of chick liver. A new cell-fractionation approach. *Biochem. J.* 138, 359–371.

Puro, D. G., and Richter, G. W. (1971). Ferritin synthesis by free and membrane-bound (poly) ribosomes of rat liver. *Proc. Soc. Exp. Biol. Med.* 138, 399–403.

Ragnotti, G., Lawford, G. R., and Campbell, P. N. (1969). Biosynthesis of microsomal nicotinamide-adenine dinucleotide phosphate-cytochrome c reductase by membrane-bound and free polysomes from rat liver. *Biochem. J.* 112, 139–147.

Randerath, K., and Randerath, E. (1973). Chemical post-labelling methods for the base composition and sequence analysis of RNA. *J. Chromatogr.* 82, 59–74.

Randerath, K. (1970). An evaluation of film detection methods for weak β-emitters, particularly tritium. *Anal. Biochem.* 34, 188–205.

Rasch, E., Riesen, A. H., and Chow, K. L. (1959) Altered structure and composition of the retinal cells in dark-reared cat. *J. Histochem. Cytochem.* 7, 321–322.

Rasch, E., Swift, H., Riesen, A. H., and Chow, K. L. (1961). Altered structure and composition of retinal cells in dark-reared mammals. *Exp. Cell. Res.* 25, 348–363.

Richardson, K., and Rose, S. P. R. (1973a). Differential incorporation of lysine into retinal protein fractions following first exposure to light. *J. Neurochem.* 21, 521–530.

Richardson, K., and Rose, S. P. R. (1973b). Differential incorporation of [^{3}H]lysine into visual cortex protein fractions during first exposure to light. *J. Neurochem.* 21, 531–537.

Riesen, A. H. (1965). Effects of early deprivation of photic stimulation. In: *The Biosocial Basis of Mental Retardation* (S. F. Osler and R. E. Cooke, eds.), pp. 61–85, Johns Hopkins Univ. Press, Baltimore.

Riesen, A. H. (1975). The sensory environment in growth and development. In: *The Developmental Neuropsychology of Sensory Deprivation* (A. H. Riesen, ed.), pp. 1–6, Academic Press, New York.

Rose, S. P. R. (1967). Preparation of enriched fractions from cerebral cortex containing isolated, metabolically active neuronal and glial cells. *Biochem. J.* 102, 33–43.

Schecter, I. (1973). Biologically and chemically pure mRNA coding for a mouse immunoglobulin L-chain prepared with the aid of antibodies and immobilized oligothymidine. *Proc. Nat. Acad. Sci. U.S.A.* 70, 2256–2260.

Sherman, S. M., Hoffman, K. P., and Stone, J. (1972). Loss of a specific cell type from dorsal lateral geniculate nucleus in visually deprived cats. *J. Neurophysiol.* 35, 532–541.

Simner, M. L. (1973). The development of visual flicker rate preference in the newly hatched chick. *Dev. Psychobiol.* 6, 377–384.

Simner, M. L. (1974). Effects of early posthatch exposure to intermittent light on visual flicker rate preference in chicks. *J. Comp. Physiol. Psychol.* 87, 267–271.

Singh, V. B., and Talwar, G. P. (1967). Effect of the flicker frequency of light and other factors on the synthesis of proteins in the occipital cortex of monkey. *J. Neurochem.* 14, 675–680.

Singh, V. B., and Talwar, G. P. (1969). Identification of a protein fraction in the occipital cortex of the monkey rapidly labelled during exposure of the animal to rhythmically flickering light. *J. Neurochem.* 16, 951–959.

Sripati, C. R., Rust, A., and Khouvine, Y. (1967). On the prevention of loss of RNA into lipid solvents. *Experientia* 23, 695–696.

Takagi, M., and Ogata, K. (1971). Isolation of serum albumin-synthesizing polysomes from rat liver. *Biochem. Biophys. Res. Commun.* 42, 125–131.

Talwar, G. P., Chopra, S. P., Goel, B. K., and D'Monte, B. (1966). Correlation of the functional activity of the brain with metabolic parameters. III. Protein metabolism of the occipital cortex in relation to light stimulus. *J. Neurochem.* 13, 109–116.

Tushmalova, N. A., and Prokofieva, L. J. (1973). Ul'trastruktura netronov gippokampa pri vyrabotke uslovnogo refleksa ukrys. Pavlov. *J. Higher Ner. Activity* 23, 651–652.

Walker, J. P., Walker, J. B., Kelley, R. L., and Riesen, A. H. (1975). Neurochemical correlates of sensory deprivation. In: *The Developmental Neuropsychology of Sensory Deprivation* (A. H. Riesen, ed.), pp. 93–124, Academic Press, New York.

Wenzel, J., David, H., Pohle, W., Marx, I, and Matties, H. (1975). Free and membrane-bound ribosomes and polysomes in hippocampal neurons during a learning experiment. *Brain Res.* 84, 99–109.

Zomzely, C. E., Roberts, S., and Peache, S. (1970). Isolation of RNA with properties of messenger RNA from cerebral polyribosomes. *Proc. Nat. Acad. Sci. U.S.A.* 67, 644–651.

Zomzely-Neurath, C., York, C., and Moore, B. W. (1973). *In vitro* synthesis of two brain-specific proteins (S100 and 14-3-2) by polyribosomes from rat brain. 1. Site of synthesis and programming by polysome-derived messenger RNA. *Arch. Biochem. Biophys.* 155, 58–69.

6

About a "Specific" Neurochemistry of Aggressive Behavior

L. VALZELLI

Istituto di Ricerche Farmacologiche "Mario Negri"
Milano, Italy

A. INTRODUCTION

In the study of brain functions, the anatomic–physiological and neurochemical correlates of emotions and behavior are important issues. It is generally accepted that patterns of animal behavior are related to the activation and mutual integration of specific neural circuits. These circuits are triggered and regulated by the availability of putative neurochemical transmitters at the synaptic sites. These mechanisms are influenced by functional needs, resulting in the anatomical and bio-chemical plasticity of the brain as modulated by environmental inputs (Bennett et al., 1964). It is possible to investigate the interaction of various elements of animal behavior as related to specific changes in brain biochemistry or, conversely, to identify induced modifications of brain biochemistry and relate them with changes in behavior.

Such correlations are easy to deduce when dramatic changes in gross behavior are associated with pronounced changes in cerebral neurotransmitters. For example, the classical postulate of Brodie and co-workers (Brodie and Shore, 1957; Brodie and Costa, 1962), suggests that behavioral sedation and activation are dependent, respectively, upon changes in brain serotonin and norepinephrine. However, the notion that a single neurotransmitter can sustain such major behavioral modifications has been disputed for years and remains questionable. It seems more difficult to advance hypotheses when alterations in single or specialized behavioral events appear to depend on changes of single cerebral metabolites.

Several factors may contribute in a negative way to the experimental data. Normal animal behavior should be considered as a "continuum," with regard to both the spontaneous or learned responses to environmental stimulations and the neurochemical or neurophysiological changes that may be involved. It is difficult to single out specific components of a given behavior and to measure and evaluate their respective differences or similarities in order to relate them to identifiable and concomitant changes of a single neurochemical substance of the whole brain. It is preferable to consider spontaneous or experimentally induced situations in which a single behavioral item becomes predominant. This situation facilitates the investigation of its biochemical and neuroanatomic correlates, especially when specific brain structures control the behavioral aspect under examination.

However the main question to investigate is whether and to what extent a single cerebral neurotransmitter is involved in the activity of selected brain structures that might directly govern a specific behavioral item. Obviously, this applies to all aspects of behavior, but, owing to the implications of the concept of violence, it seems to have particular relevance when aggressiveness is examined.

B. BIOCHEMICAL CONCOMITANTS OF AGGRESSIVENESS BY ISOLATION

Regarding prolonged isolation or socioenvironmental deprivation, it has been shown that social interaction and the mean level of environmental stimulation (Welch, 1965) are essential factors for shaping anatomic, neurochemical, affective, and behavioral characteristics of animals (Bennett et al., 1969; Denenberg et al., 1964; Diamond et al., 1964, 1966; Essman, 1966, 1968, 1969; Ferchmin et al., 1970; Rosenzweig, 1966; Rosenzweig et al., 1968; Valzelli, 1966, 1967, 1971, 1973a, 1976; Valzelli and Garattini, 1968, 1972; Welch and Welch, 1969; Yen et al., 1959). Prolonged isolation in adult laboratory animals induces a strong, repetitive, and compulsive inter- and/or intraspecies and intermale

Table I. Time Required to Achieve a 50% Increase in the
Basal Value of Brain Serotonin (5-HT)
after MAO blockade[a]

		Time (min)	
Drug	Dose (mg/kg)	G	A
Tranylcypromine	20	22	36[b]
Pheniprazine	20	27	43[b]
Pargyline	100	28	73[b]

[a] Normally grouped (G) and isolated-aggressive (A) male Swiss Albino
mice were injected intraperitonially with different drugs.
[b] $P < 0.01$.

aggressive behavior (Allee, 1942; Scott, 1958; Seward, 1946; Valzelli, 1966, 1967; Yen et al., 1959) which is sex and strain dependent (Bevan et al., 1951; Valzelli, 1971; Valzelli and Garattini, 1968, 1972). As reported elsewhere (Valzelli, 1973a), such aggressive behavior, which is the most evident effect of prolonged isolation, may represent a behavior abnormality since it appears to be completely aimless and is associated with a series of other somatic, behavioral, and neurochemical changes (Essman, 1968, 1969, 1971, 1974; Garattini et al., 1969; Valzelli, 1971, 1973a,b, 1974; Valzelli and Garattini, 1972; Welch and Welch, 1969, 1971). The earlier observation (Valzelli, 1966) suggesting that isolation-induced aggressiveness is accompanied by cerebral biochemical changes was based on the finding that aggressive mice tended to show an increase in brain serotonin after monoamine oxidase blockade which differed in the time of onset from that of normal mice (Table I). However, it was also pointed out that this biochemical change was not necessarily responsible for the aggressiveness and could have been a concomitant event of the isolation procedure (Valzelli, 1966).

Concerning the role played by brain monoamines in isolation-induced aggressiveness, two main views postulate that changes in either brain catecholamines or serotonin are responsible for this behavioral pattern (Eichelman, 1973; Hodge and Butcher, 1974).

It has been observed that the increase of brain serotonin that follows administration of 5-hydroxytryptophan, monoamine oxidase inhibitors, or lysergide reduces isolation-induced aggressiveness in mice and muricidal behavior in rats (Bocknik and Kulkarni, 1974; Kulkarni, 1968a; Hodge and Butcher, 1974; Sheard, 1969; Rewerski et al., 1971; Uyeno and Benson, 1965; Valzelli et al., 1967; Welch and Welch, 1968, Yen et al.,

1959). Conversely, a decrement in serotonergic function has been correlated with increased aggression. Koe and Weissman (1966) observed that normal rats injected with p-chlorophenylalanine displayed increased irritability and sometimes overt aggressiveness when handled. Administration of this drug also increased muricidal behavior (Sheard, 1969) and accelerated and potentiated the isolation-induced muricidal behavior in previously nonmuricidal rats (Di Chiara et al., 1971). Olfactory bulbectomy induced muricidal behavior in nonkiller rats and lowered the serotonin concentration of the amygdala (Karli et al., 1969, 1972). A consistent decrease in the level of this amine occurred also after lesioning the midbrain raphé nuclei (Vergnes et al., 1974). Such a lesion induced a variable muricidal aggression in normal rats (Banerjee, 1974; Grant et al., 1973; Vergnes et al., 1974), whereas the selective lowering of brain serotonin, but not of brain catecholamines, increased shock-elicited fighting in mice (Butcher and Dietrich, 1973). The same result was obtained in isolated mice when the activity of 5-hydroxytryptophan decarboxylase was inhibited (Hodge and Butcher, 1974).

Interestingly, when naive mice were exposed to repeated aggression and defeat by highly aggressive mice, their cerebral 5-hydroxytryptophan decarboxylase activity was consistently decreased in the amygdala and consistently increased in the frontal cortex (Eleftheriou and Church, 1968). Therefore, in accord with the results of Hodge and Butcher (1974), the role of serotoninergic brain activity in aggression appears to be well correlated with the inhibitory control of such behavior, as balanced with the activity of other biological principles.

Therefore, the biochemical results described, as induced by prolonged isolation with regard to aggressive behavior, may be as well probative. If it is assumed that changes in brain monoamine turnover correlate to some extent with changes in their concentration, we must take into consideration the findings that prolonged isolation does not alter brain monoamine levels in the whole brain or in selected brain areas of male mice (Garattini et al., 1967, 1969; Valzelli, 1966), whereas there are decreases in brain serotonin turnover in the whole brain, diencephalon, and corpora quadrigemina (Garattini et al., 1969; Valzelli, 1971) (Tables 11 and III).

It has been shown that the decrease in serotonin turnover begins after 24 hr of isolation and remains significant throughout the 30-day period of isolation (Table IV). It did not occur in female mice and in strains of mice and rats which did not become aggressive during socioenvironmental deprivation (Tables V, VI, and VII). This change in serotonin turnover in mice was accompanied by a slight decrease in brain norepinephrine turnover and by a substantial increase in brain

Table II. Brain Serotonin (5-HT), Norepinephrine (NE), and Dopamine (DA) Contents in Whole Brains and in Various Brain Areas[a]

Brain area	Weight %	5-HT		NE		DA	
		N	A	N	A	N	A
Whole brain	100	0.65 ± 0.03	0.65 ± 0.03	0.45 ± 0.02	0.42 ± 0.03	1.09 ± 0.05	1.04 ± 0.04
Cerebellum	15	—	—	0.10 ± 0.02	0.10 ± 0.01	—	—
Olfactory bulbs	5	0.39 ± 0.01	0.40 ± 0.06	0.11 ± 0.02	0.12 ± 0.04	0.68 ± 0.02	0.73 ± 0.03
Quadrigeminal bodies	5	1.10 ± 0.07	1.01 ± 0.04	0.60 ± 0.05	0.49 ± 0.03	0.71 ± 0.02	0.67 ± 0.02
Midbrain	12	0.62 ± 0.03	0.58 ± 0.04	0.39 ± 0.05	0.36 ± 0.03	0.06 ± 0.005	0.04 ± 0.002
Diencephalon	13	0.94 ± 0.02	0.94 ± 0.03	0.55 ± 0.05	0.50 ± 0.04	0.29 ± 0.01	0.28 ± 0.02
Rest of the brain	50	0.40 ± 0.03	0.47 ± 0.02	0.36 ± 0.03	0.29 ± 0.04	1.77 ± 0.01	1.60 ± 0.02

[a] A, isolated-aggressive male Swiss Albino mice; B, normal male Swiss Albino mice.

Table III. Brain Serotonin (5-HT) Turnover in Various Brain Areas[a]

| | Brain 5-HT turnover | | | |
| | Rate (μg/g/hr) | | Time (min) | |
Brain area	G	A	G	A
Whole brain	0.45	0.30[b]	90	130[b]
Hemispheres	0.29	0.25	120	131[c]
Diencephalon	0.95	0.73[b]	70	98[b]
Mesencephalon	0.56	0.51	86	94
Quadrigeminal bodies	0.80	0.52[c]	77	90[c]

[a] G, normally grouped male Swiss Albino mice; A, isolated-aggressive male Swiss Albino mice.
[b] $P < 0.01$.
[c] $P < 0.05$.

dopamine turnover (Valzelli, 1971; see Table VIII). The hypothesis that aggressiveness in mice might be conditioned by an enhanced dopamine synthesis and a simultaneous reduction of serotonin formation, or by a change in the balance between serotonin and dopamine in favor of the latter neurotransmitter, has been previously formulated by Lycke et al. (1969) who observed an increase of dopamine with a concomitant decrease of serotonin in the brains of aggressive encephalitic mice infected by *Herpes simplex* virus. A similar conclusion has been reached by others with respect to the fighting behavior and other behavioral changes induced by multiple drug treatment (Benkert et al., 1973a,b).

Prolonged socioenvironmental deprivation is known to produce muricidal behavior in a certain percentage of male rats of some strains (Goldberg and Salama, 1969; Kulkarni, 1968b; Myer, 1964; Valzelli and Garattini, 1972), and to induce changes in cerebral monoamine turnover in muricidal animals. Muricidal rats exhibit an increased turnover rate of brain norepinephrine (Goldberg and Salama, 1969; Valzelli, 1971; Valzelli and Garattini, 1972) that occurs concomitantly with a decreased turnover of brain serotonin (Valzelli, 1971; Valzelli and Garattini, 1972). These findings are consistent with the demonstration that prolonged isolation induces an increase of tyrosine hydroxylase activity and a decrease in tryptophan hydroxylase activity of rat brain (Segal et al., 1973). However, in only partial agreement with previous findings (Valzelli, 1971), other studies in mice have indicated that a reduction in norepinephrine and an increase in dopamine turnover occurred in the brain without any consistent change in serotonin turn-

Table IV. Development of Changes in Brain Serotonin (5-HT) Turnover and in Aggressive Behavior in Male Swiss Albino Mice during Isolation

No. of animals per cage	Days of isolation	5-HT turnover		% of aggressive animals
		Rate (μg/g/hr)	Time (min)	
1	1	0.29 ± 0.01[a]	144 ± 7	15
	5	0.32 ± 0.02[a]	140 ± 2	20
	10	0.30 ± 0.02[a]	124 ± 4	38
	20	0.28 ± 0.01[a]	140 ± 3	86
	30	0.26 ± 0.01[a]	151 ± 2	100
10	30	0.46 ± 0.01	80 ± 3	0

[a] $p < 0.001$.

over (Modigh, 1973, 1974). Intensive fighting, as a cause of a stressful situation, induces a rapid acceleration of the turnover of the three monoamines (Modigh, 1973). Aggressive behavior and brain monoamines were considered unrelated if there was no change or a decrease in brain norepinephrine turnover, without any modification in serotonin turnover, which may occur following prolonged socioenvironmental deprivation of different strains of mice (Goldberg et al., 1973). Conflicting biochemical changes have been induced by olfactory bulbectomy in male rats: a decrease in the serotonin content of the amygdala has been found (Karli et al., 1969), where as lowered cortical brain norepinephrine has also been reported without alteration of dopamine or serotonin levels, regardless of whether the animals became muricidal (Eichelman et al., 1972a). Depletions of brain dopamine and norepinephrine following degeneration of catecholaminergic brain terminals induced by 6-hydroxydopamine administration facilitated shock-elicited aggression in rats (Eichelman et al., 1972b). In contrast, the degeneration of brain serotonergic terminals induced by 5,6-dihydroxytryptamine (Baumgarten et al., 1972a,b,c) produced muricidal behavior in rats (Breese et al., 1974), indicating that the relationship between aggressive behavior and changes in brain monoamines remains unresolved.

The most obvious explanations for these conflicting findings are: (a) Divers techniques are used to investigate different aspects of aggressive behavior in varied species and animal strains; (b) changes in monoamine turnover are frequently measured with an assortment of techniques in brain areas which are not comparable, and these changes

Table V. Brain Serotonin (5-HT) Turnover and Aggressive Behavior in Male and Female Swiss Albino Mice after Prolonged Isolation[a]

| Sex | Days of isola- tion | 5-HT turnover | | | | % of aggressive animals | |
| | | Rate (μg/g/hr) | | Time (min) | | | |
		I	G	I	G	I	G
Male	1	0.30 ± 0.02[b]	0.46 ± 0.03	132 ± 10	84 ± 8	15	0
	30	0.29 ± 0.02[b]	0.46 ± 0.02	132 ± 10	84 ± 8	100	0
Female	1	0.42 ± 0.05	0.42 ± 0.01	101 ± 12	105 ± 12	0	0
	30	0.42 ± 0.04	0.47 ± 0.01	94 ± 10	81 ± 3	0	0

[a] I, isolated animals; G, grouped animals.
[b] $p < 0.001$.

in turnover may derive only from alterations in the levels of some biochemical components; (c) other possibly related components, such as cholinergic mechanisms (Allikmets, 1974; Karli and Mandel, 1974) are seldom considered; (d) different kinds of aggression (see Moyer, 1968) may have different biochemical correlates (Eichelman and Thoa, 1973); and (e) as discussed earlier, the emotional baseline of the experimental animal underlying differential monoamine availability and monoaminergic activity may profoundly affect both behavioral and biochemical results of any procedure. Olfactory bulbectomy, which induces muricidal behavior in previously normal rats, can abolish aggression in spontaneously aggressive animals (Denenberg et al., 1973; Ropartz, 1968). Furthermore, unpublished studies in our laboratory have shown that lesioning of the midbrain raphé nuclei in male mice, before subjecting the animals to prolonged socioenvironmental deprivation accelerated the onset of aggressive behavior. Kostowski and Valzelli (1974) also reported that this intervention, performed in isolation-induced aggressive mice, completely abolished their aggression (Fig. 1).

C. NEUROANATOMIC CONSIDERATIONS

Another significant aspect of this problem that should be considered involves the functional neuroanatomic substrates responsible for specific behavioral changes. It must be remembered that a neurotransmitter, although involved in the function of neuroanatomical systems, cannot in itself determine the nature of behavioral responses. The importance

Table VI. Effect of Prolonged Isolation upon Brain Serotonin (5-HT) Turnover and Behavior of Different Mouse Strains[a]

	Brain 5-HT turnover				
	Rate (μg/g/hr)		Time (min)		Intensity of
Strain of mice	G	I	G	I	aggressiveness
Albino Swiss	0.45	0.28[b]	90	130[b]	100
C_3H	0.85	0.59[b]	45	60[b]	100
DBA	0.61	0.52	70	57	50
CBA/J	0.63	0.70	80	74	0
$C_{57}B1/J6$	0.67	0.60	68	73	0
BALB/C	0.34	0.34	103	96	0
BDF/1	0.50	0.45	54	60	0

[a] I, isolated animals; G, grouped animals.
[b] $P < 0.01$.

of the limbic system in governing animal behavior derives from its manifold associative and integrative functions. From the functional standpoint, the entire limbic system should be regarded as a "continuum," and the activation or inhibition of one of its structures can initiate a multilink sequence that involves most, if not all, brain structures. This may explain why the mechanism for the most elementary patterns of behavior may increase its complexity along with the development of the brain. Within limits of such considerations, a series, of behavioral events can be associated with some identifiable cerebral structures.

The hypothalamus is known to play an important role in feeding behavior which can also be associated with aggression. Stimulation of the lateral hypothalamus may cause feeding behavior, and augment food intake (Delgado and Anand, 1953; Miller, 1957, 1961; Oomura et al., 1967; Smith, 1956), and predatory aggressiveness may occur in the presence of an appropriate attack object (Wasman and Flynn, 1962). Stimulation of the medial hypothalamus produces a different kind of aggression, in that a cat may ignore the presence of an available rat but may viciously attack the experimenter (Egger and Flynn, 1963).

Lesions of the medial hypothalamus enhance the defense behavior of rats in a shock-box, whereas lesions confined to lateral hypothalamus can abolish "territorial" aggression (Adams, 1971). Bilateral lesions of the amygdala tame a variety of innately hostile and vicious animals (Schreiner and Kling, 1956; Woods, 1956) and block interspecific, com-

Table VII. Effect of Prolonged Isolation on Brain Serotonin (5-HT)
Turnover and Behavior of Different Rat Strains[a]

| | Brain 5-HT turnover | | | | Muricidal behavior (% of isolated animals) |
| | Rate (μg/g/hr) | | Time (min) | | |
Rat strain	G	I	G	I	
Sprague-Dawley	0.38	0.36	70	71	—
Buffalo	0.37	0.35	70	71	—
Wistar	0.33	0.25[b]	72	88[b]	38

[a] G, grouped animals; I, isolated animals.
[b] $P < 0.01$.

petitive aggression and muricidal behavior in laboratory rats (Karli, 1974; Karli and Vergnes, 1964a,b; Vergnes and Karli, 1964, 1965). However, bilateral destruction of either the basal or central nuclei of the amygdala produces the opposite result when performed in already domesticated and friendly animals which, after the intervention, attack without provocation (Wood, 1958). A similar bimodal effect has been obtained after olfactory bulb removal which induces muricidal behavior, violent intermale aggression, and rat pup-killing in previously normal laboratory rats (Didiergeorges et al., 1966; Karli et al., 1969; Myer, 1964), and instead eliminates spontaneous or natural frog-killing, mouse-killing, and intermale aggression in rats (Bandler and Chi, 1972) as well as intermale aggression in naturally vicious mice and hamsters (Murphy, 1970; Ropartz, 1968; Rowe and Edwards, 1971). Therefore, it is of interest to speculate whether and to what extent the emotional baseline might modulate the final outcome of such a surgical intervention in brain, or if the different results obtained are merely dependent on the location specificity, and size of the lesion. Several other brain areas are also known to be involved in the modulation of irritative, hostile, or overt aggressive behavior. For example, lesions of the ventro-medial hypothalamus produce extreme viciousness in cats (Wheatley, 1944), and increased aggression results from destruction of the septal area, frontal lobes, cingulum, portions of hippocampus, and ventral midbrain tegmentum (Bandler et al., 1972; Brady and Nauta, 1955; Heller et al., 1962; Karli, 1955, 1956; Kenyon and Krieckhaus, 1965; Moyer, 1968, 1969; Vergnes and Karli, 1968; Zeman and Innes, 1963).

Radio stimulation of the posterior ventrolateral nucleus of the thalamus or of the central gray matter has been shown to evoke offensive–

Table VIII. Effect of Prolonged Isolation of Male Swiss Albino Mice
on Brain Monoamine Turnover

| | Brain monoamine turnover | | | | | |
| | Serotonin | | Norepinephrine | | Dopamine | |
Type of mice	Rate (μg/g/hr)	Time (min)	Rate (μg/g/hr)	Time (hr)	Rate (μg/g/hr)	Time (hr)
Normal	0.49	74	0.07	5.67	0.08	6.17
Aggressive	0.26[a]	150[a]	0.03[a]	8.58[a]	0.15[a]	3.50[a]

[a] $P < 0.01$.

defensive patterns in monkey colonies (Delgado, 1967). Spontaneous aggressive behavior of monkeys can be inhibited by radio stimulation of the head of the caudate nucleus, a procedure which can also arrest in full charge animals as dangerous as brave bulls. Furthermore, intermittently programmed stimulation of caudate nucleus sites may inhibit a dominant animal, thereby changing the hierarchical structure of the entire social group (Delgado, 1974). The fimbria of the fornix also participates in the offensive–defensive system and with the posterior hippocampus, is involved in the integration of nociceptive responses (Delgado, 1955).

D. CONCLUSIONS

When neurochemical correlates of behavior are examined, an attempt should be made to evaluate possible changes in the absolute level and/or in the turnover of biochemically active principles to single brain areas or to a series of single and precisely identified structures which appear to mediate given behavioral patterns. It should be recognized that in addition to influencing each other, serotonin, norepinephrine, and dopamine also interact with the functions and metabolism of several other putative neurotransmitters. Regarding serotonin, its relations with catecholamines in the central nervous system have recently been demonstrated (Blondaux et al., 1973 Héry et al., 1973; Kostowski et al., 1974; Johnson et al., 1972; Jouvet, 1973; Lichtensteiger et al., 1967).

In most cases, a multiplicity of nuclei exist within the same brain structure, and their functions may be partially or completely synergistic or agonistic. These regions perhaps require different neurotransmitters

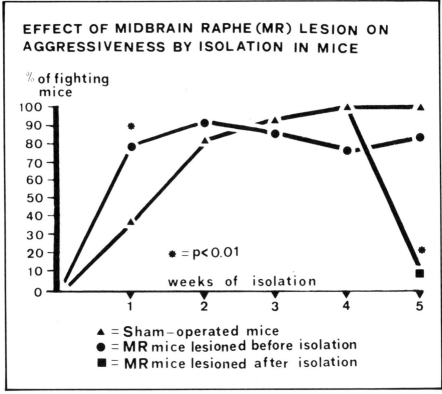

EFFECT OF MIDBRAIN RAPHE (MR) LESION ON AGGRESSIVENESS BY ISOLATION IN MICE

% of fighting mice

* = p<0.01

weeks of isolation

▲ = Sham–operated mice
● = MR mice lesioned before isolation
■ = MR mice lesioned after isolation

1 Effect of a midbrain raphe (MR) lesion on aggressiveness by isolation of mice. △, Sham-operated mice; ●, MR mice lesioned before isolation; ■, MR mice lesioned after operation.

for their normal function. To consider structures such as the hypothalamus and amygdala anatomically homogeneous is obviously incorrect, especially in regard to behavioral expression in general and to aggressiveness in particular. Cerebral structures contain a number of different nuclei, the combined activities of which determine moment by moment the graded intensity of functional activity sometimes attributed to whole structures, which are balanced by the activity of other cerebral areas.

This complexity is also reflected by the presence of different neurochemical substrates in the same structure. For example, acetylcholine, serotonin, norepinephrine, and dopamine are all present in the amygdala, thus allowing for an interrelated biochemistry which subserves the multiple and integrated functional activity of this structure. To speak in terms of inhibiting or activating neurotransmitters may be misleading, since both inhibiting and activating circuits and systems are present in the brain and usually function by means of a given neurotransmitter.

Also, the biochemical significance of selective procedures employed for suppressing or activating a monoaminergic brain region may be misinterpreted in relation to given behavioral responses. It may still be acceptable to consider a given neurotransmitter involved in sustaining large "segments" of general behavior such as sleep or wakefulness, gross excitation or gross sedation, general arousal, or general inactivation.

The change of a given neurotransmitter should be taken only as an indication that the entire machinery or region of the brain (serotonergic or catecholaminergic) subserved by this neurotransmitter is wrongly tuned. This incorrect tuning may involve the functioning of several different structures and circuitries thereby disrupting their reciprocal balance and leading to a series of disturbed behaviors rather than compromising selectively a single behavioral element. Therefore, to speak in terms of a single neurochemical transmitter or of a "specific" neurochemistry as being responsible for aggressiveness or for any other behavioral element seems to be inappropriate and probably misleading, especially since aggressiveness is not a single entity (Valzelli, 1967; Moyer, 1968).

E. REFERENCES

Adams, D. B. (1971). Defense and territorial behaviour dissociated by hypothalamic lesions in the rat. *Nature* (*London*) 232, 573–574.

Allee, W. C. (1942). Social dominance and subordination among vertebrates. *Biol. Symp.* 8, 139–145.

Allikmets, L. H. (1974). Cholinergic mechanisms in aggressive behavior. *Med. Biol.* 52, 19–30.

Bandler, R. J., and Chi, C. C. (1972). Effects of olfactory bulb removal on aggression: A reevaluation. *Physiol. Behav.* 8, 207–211.

Bandler, R. J. Jr., Chi, C. C., and Flynn, J. P. (1972). Biting attack elicited by stimulation of the ventral midbrain tegmentum of cats. *Science* 177, 364–366.

Banerjee, U. (1974). Modification of the isolation-induced abnormal behavior in male Wistar rats by destructive manipulation of the central monoaminergic systems. *Behav. Biol.* 11, 573–579.

Baumgarten, H. G., Björklund, A., Lachenmayer, L., Nobin, A., and Stenevi, U. (1972a). Long-lasting selective depletion of brain serotonin by 5,6-dihydroxytryptamine. *Acta Physiol. Scand.* 84 (Suppl. 373), 1–15.

Baumgarten, H. G., Björklund, A., Holstein, A. F., and Nobin, A. (1972b).

Chemical degeneration of indoleamine axons in the rat brain by 5-6-di-hydroxytryptamine: An ultrastructural study. *Z. Zellforsch. Mikrosk. Anat.* 129, 256–271.

Baumgarten, H. G., Evetts, K. D., Holman, R. B., Iversen, L. L., Vogt. M., and Wilson, G. (1972c). Effects of 5,6-dihydroxytryptamine on monoaminergic neurones in the central nervous system of the rat. *J. Neurochem.* 19, 1587–1597.

Benkert, O., Gluba, H., and Mattussek, N. (1973a). Dopamine, noradrenaline and 5-hydroxytryptamine in relation to motor activity, fighting and mounting behaviour. I. L-DOPA and D,L-threo-dihydroxyphenylserine in combination with RO-4-4602, pargyline and reserpine. *Neuropharmacology* 12, 177–186.

Benkert, O., Renz, A., and Matussek, N. (1973b). Dopamine, noradrenaline and 5-hydroxytryptamine in relation to motor activity, fighting and mounting behaviour. II. L-DOPA and D,L-threo-dihydroxyphenylserine in combination with RO-4-4602 and parachlorophenylalanine. *Neuropharmacology* 12, 187–193.

Bennett, E. L., Diamond, M. C., Krech, D. and Rosenzweig, M. R. (1964). Chemical and anatomical plasticity of brain. *Science* 146, 610–619.

Bennett, E. L., Rosenzweig, M. R., and Diamond, M. C. (1969). Rat brain: Effects of environmental enrichment on wet and dry weights. *Science* 163, 825–826.

Bevan, W., Jr., Bloom, W. L., and Lewis, G. T. (1951). *Physiol. Zool.* 24, 231.

Blondaux, C., Juge, A., Sordet, F., Chouvet, G., Jouvet, M., and Pujol, J.-F. (1973). Modification du métabolisme de la sérotonine (5HT) cérébrale indiute chez le rat par administration de 6-hydroxydopamine. *Brain Res.* 50, 101–114.

Bocknik, S. E., and Kulkarni, A. S. (1974). Effect of a decarboxylase inhibitor (RO-4-4602) on 5-HTP induced muricide blockade in rats. *Neuropharmacology* 13, 279–281.

Brady, J. V., and Nauta, W. J. H. (1955). Subcortical mechanisms in emotional behavior: the duration of affective changes following septal and habenular lesions in the rat. *J. Comp. Physiol. Psychol.* 48, 412–420.

Breese, G. R., Cooper, B. R., Grant, L. D., and Smith, R. D. (1974). Biochemical and behavioural alterations following 5-6-dihydroxytryptamine administration into brain. *Neuropharmacology* 13, 177–187.

Brodie, B. B., and Costa, E. (1962). Some current views on brain monoamines. In: *Monoamines et Système Nerveux Central*, (J. de Ajuriaguerra, ed.), pp. 13–49, George & Cie, Genève.

Brodie, B. B., and Shore, P. A. (1957). A concept for a role of serotonin and norepinephrine as chemical mediators in the brain. *Ann. N. Y. Acad. Sci. U.S.A.* 66, 631–642.

Butcher, L. L., and Dietrich, A. P. (1973). Effects of shock-elicited aggression in mice of preferentially protecting brain monoamines against the depleting action of reserpine. *Naunyn Schmiedebergs Arch. Pharmacol.* 277, 61–70.

Delgado, J. M. R. (1955). Cerebral structures involved in transmission and elaboration of noxious stimulation. *J. Neurophysiol.* 18, 261–275.

Delgado, J. M. R. (1967). Limbic system and free behavior. In: *Progress in Brain Research: Structure and Functions of the Limbic System.* (W. R. Adey and T. Tokizane, eds.) Vol. 27, pp. 48–68, Elsevier, Amsterdam.

Delgado, J. M. R. (1974). Communication with the conscious brain by means of electrical and chemical probes. In: *Factors in Depression* (N. S. Kline, ed.), pp. 251–268, Raven Press, New York.

Delgado, J. M. R., and Anand, B. K. (1953). Increase of food intake induced by electrical stimulation of the lateral hypothalamus. *Amer. J. Physiol.* 172, 162–168.

Denenberg, V. H., Gaulin-Kremer, E., Gandelman, R., and Zarrow, M. X. (1973). The development of standard stimulus animals for mouse (*Mus musculus*) aggression testing by means of olfactory bulbectomy. *Anim. Behav.* 21, 590–598.

Denenberg, V. H., Morton, J. R., and Haltmeyer, G. C. (1964). Effect of social grouping upon emotional behaviour. *Anim. Behav.* 12, 205–208.

Diamond, M. C., Krech, D., and Rosenzweig, M. R. (1964). The effects of an enriched environment on the histology of the rat cerebral cortex. *J. Comp. Neurol.* 123, 111–120.

Diamond, M. C., Law, F., Rhodes, H., Lindner, B., Rosenzweig, M. R., Krech, D., and Bennett, E. L. (1966) Increases in cortical depth and glia numbers in rats subjected to enriched environment. *J. Comp. Neurol.* 128, 117–126.

Di Chiara, G., Camba, R., and Spano, P. F. (1971). Evidence for inhibition by brain serotonin of mouse killing behavior in rats. *Nature (London)* 233, 272–273.

Didiergeorges, F., Vergnes, M., and Karli, P. (1966). Privation des afférences olfactives et agressivité interspécifique du rat. *C. R. Seanc. Soc. Biol.* 160, 866–868.

Egger, M. D., and Flynn, J. P. (1963). Effect of electrical stimulation of the amygdala and hypothalamically elicited attack behavior in cats. *J. Neurophysiol.* 26, 705–720.

Eichelman, B. (1973). The catecholamines and aggressive behavior. In: *Neurosciences Research: Chemical Approaches to Brain Function* (S. Ehrenpreis and I. J. Kopin, eds.), Vol. 5, pp. 109–129, Academic Press, New York.

Eichelman, B. S., Jr., and Thoa, N. B. (1973). The aggressive monoamines. *Biol. Psychiat.* 6, 143–164.

Eichelman, B. S., Jr., Thoa, N. B., and Ng, K. Y. (1972a). Facilitated aggression in the rat following 6-hydroxydopamine administration. *Physiol. Behav.* 8, 1–3.

Eichelman, B., Thoa, N. B., Bugbee, N. M., and Ng, K. Y. (1972b). Brain amine and adrenal enzyme levels in aggressive bulbectomized rats. *Physiol. Behav.* 9, 483–485.

Eleftheriou, B. E., and Church, R. L. (1968). Brain 5-hydroxytryptophan decarboxylase in mice after exposure to aggression and defeat. *Physiol. Behav.* 3, 323–325.

Essman, W. B., (1966). The development of activity differences in isolated and aggregated mice. *Anim. Behav.* 14, 406–409.

Essman, W. B. (1968). Differences in locomotor activity and brain-serotonin metabolism in differentially housed mice. *J. Comp. Physiol. Psychol.* 66, 244–246.

Essman, W. B. (1969). "Free" and motivated behaviour and amine metabolism in isolated mice. In: *Aggressive Behaviour* (S. Garattini and E. B. Sigg, eds.), pp. 203–208, Excerpta Medica, Amsterdam.

Essman, W. B. (1971). Neurochemical changes associated with isolation and environmental stimulation. *Biol. Psychiat.* 3, 141–147.

Essman, W. B. (1974). Regional alterations of synaptic o-phosphorylethanolamine in differentially housed mice. *Pharmacol. Res. Commun.* 6, 377–395.

Ferchmin, P. A., Eterovic, V. A., and Caputto, R. (1970). Studies on brain weight and RNA content after short periods of exposure to environmental complexity. *Brain Res.* 20, 49–57.

Garattini, S., Giacalone, E., and Valzelli, L. (1967). Isolation, aggressiveness and brain 5-hydroxytryptamine turnover. *J. Pharm. Pharmacol.* 19, 338–339.

Garattini, S., Giacalone, E., and Valzelli, L. (1969). Biochemical changes during isolation-induced aggressiveness in mice. In: *Aggressive Behaviour* (S. Garattini and E. B. Sigg, eds.), pp. 179–187, Excerpta Medica, Amsterdam.

Goldberg, M. E., Insalaco, J. R., Hefner, M. A., and Salama, A. I. (1973). Effect of prolonged isolation on learning, biogenic amine turnover and aggressive behaviour in three strains of mice. *Neuropharmacology* 12, 1049–1058.

Goldberg, M. E., and Salama, A. I. (1969). Norepinephrine turnover and brain monoamine levels in aggressive mouse-killing rats. *Biochem. Pharmacol.* 18, 532–534.

Grant, L. D., Coscina, D. V., Grossman, S. P., and Freedman, D. X. (1973). Muricide after serotonin depleting lesions of midbrain raphé nuclei. *Pharmacol. Biochem. Behav.* 1, 77–80.

Heller, A., Harvey, J. A., and Moore, R. Y. (1962). A demonstration of a fall in brain serotonin following central nervous system lesions in the rat. *Biochem. Pharmacol.* 11, 859–866.

Héry, F., Rouer, E., and Glowinski, J. (1973). Effect of 6-hydroxydopamine on daily variations of 5-HT synthesis in the hypothalamus of the rat. *Brain Res.* 58, 135–146.

Hodge, G. K., and Butcher, L. L. (1974). 5-Hydroxytryptamine correlates of isolation-induced aggression in mice. *Eur. J. Pharmacol.* 28, 326–337.

Johnson, D. N., Funderburk, W. H., Ruckart, R. T., and Ward, J. W. (1972). Contrasting effect of two 5-hydroxytryptamine-depleting drugs on sleep patterns in cats. *Eur. J. Pharmacol.* 20, 80–84.

Jouvet, M. (1973). Monoaminergic regulation of the sleep-waking cycle. In: *Pharmacology in the Future of Man* (G. H. Acheson, ed.), Vol. 4, pp. 103–107, Karger, Basel.

Karli, P. (1955). Effets de lésions experimentales des noyaux amygdaliens et du lobe frontal sur le comportement d'agression du rat vis-à-vis de la souris. *C. R. Seanc. Soc. Biol.* 149, 2227–2229.

Karli, P. (1956). The Norway rat's killing response to the white mouse. An experimental analysis. *Behaviour* 10, 81–103.

Karli, P. (1974). Aggressive behavior and its brain mechanisms (as exemplified by an experimental analysis of the rat's mouse-killing behavior). In: *Determinants and Origins of Aggressive Behaviour* (J. de Wit and W. W. Hartup, eds.), pp. 277–290, Mouton, The Hague.

Karli, P., and Mandel, P. (1974). Amygdala and aggressiveness in rodents: Neurochemical correlates. *J. Pharmacol. (Paris)* 5, 91–92.

Karli, P., and Vergnes, M. (1964a). Nouvelles données sur les bases neurophysiologiques du comportement d'agression intérspecifiques rat-souris. *J. Physiol. (Paris)* 56, 384.

Karli, P., and Vergnes, M. (1964b). Dissociation expérimentale du comportement d'agression interpsécifique rat-souris et du comportement alimentaire. *C. R. Seanc. Soc. Biol.* 158, 650–653.

Karli, P., Vergnes, M., and Didiergeorges, F. (1969). Rat-mouse interspecific aggressive behaviour and its manipulation by brain ablation and brain stimulation. In: *Aggressive Behaviour* (S. Garattini, and E. B. Sigg, eds.), pp. 47–55, Excerpta Medica, Amsterdam.

Karli, P., Vergnes, M., Eclancher, F., Scmitt, P., and Chaurand, J. P. (1972). Role of the amygdala in the control of "mouse-killing" behavior in the rat. In: *The Neurobiology of the Amygdala: Advances in Behavioral Biology* (B. E. Eleftheriou, ed.), Vol. 2, pp. 553–580, Plenum Press, New York.

Kenyon, J., and Krieckhaus, E. E. (1965). Enhanced avoidance behavior following septal lesions in the rat as a function of lesion size and spontaneous activity. *J. Comp. Physiol. Psychol.* 59, 466–469.

Koe, B., and Weissman, A. (1966). *p*-Chlorophenylalanine: A specific depletor of brain serotonin. *J. Pharmacol. Exp. Therapeut.* 154, 499–516.

Kostowski, W., Samanin, R., Bareggi, S. R., Marc, V., Garattini, S., and Valzelli, L. (1974). Biochemical aspects of the interaction between midbrain raphé and locus coeruleus in the rat. *Brain Res.* 82, 178–182.

Kostowski, W., and Valzelli, L. (1974). Biochemical and behavioral effects of lesions of raphé nuclei in aggressive mice. *Pharmacol. Biochem. Behav.* 2, 277–280.

Kulkarni, A. S. (1968a). Muricidal block produced by 5-hydroxytryptophan and various drugs. *Life Sci.* 7, 125–128.

Kulkarni, A. S. (1968b). Satiation of instinctive mouse killing by rats. *Psychol. Rec.* 18, 385–388.

Lichtensteiger, W., Mutzner, U., and Langemann, H. (1967). Uptake of 5-hydroxytryptamine and 5-hydroxytryptophan by neurons of the central nervous system normally containing catecholamines. *J. Neurochem.* 14, 489–497.

Lycke, E., Modigh, K., and Roos, B.-E. (1969). Aggression in mice associated with changes in the monoamine-metabolism of the brain. *Experientia* 25, 951–953.

Miller, N. E. (1957). Experiments on motivation. *Science* 126, 1271–1278.

Miller, N. E. (1961). Implications for theories of reinforcement. In: *Electrical Stimulation of the Brain* (D. E. Sheer, ed.), pp. 515-581, University of Texas Press, Austin.

Modigh, K. (1973). Effects of isolation and fighting in mice on the rate of synthesis of noradrenaline, dopamine and 5-hydroxytryptamine in the brain. *Psychopharmacologia* 33, 1–17.

Modigh, K. (1974). Effect of social stress on the turnover of brain catecholamines and 5-hydroxyptryptamine in mice. *Acta Pharmacol. Toxicol.* 34, 97–105.

Moyer, K. E. (1968). Kinds of aggression and their physiological basis. *Commun. Behav. Biol.* 2, part A, 65–87.

Moyer, K. E. (1969). Internal impulses to aggression. *Trans. New York Acad. Sci.* 31, 104–114.

Murphy, M. R. (1970). Olfactory bulb removal reduces social and territorial behaviors in the male golden hamster. Paper presented at Eastern Psychological Association.

Myer, J. S. (1964). Stimulus control of mouse-killing rats. *J. Comp. Physiol. Psychol.* 58, 112–117.

Oomura, Y., Ooyama, H., Yamamoto, T., Naka, F., Kobayashi, N., and Ono, T. (1967). Neuronal mechanism in feeding. In: *Progress in Brain Research: Structure and Function of the Limbic System* (W. R. Adey, and T. Tokizane, eds.), Vol. 27, pp. 1–33, Elsevier, Amsterdam.

Rewerski, W., Kostowski, W., Piechcki, T., and Rylski, M. (1971). The effect of some hallucinogens on aggressiveness of mice and rats. *Pharmacology* 5, 314–320.

Ropartz, P. (1968). The relation between olfactory stimulation and aggressive behaviour in mice. *Anim. Behav.* 16, 97–100.

Rosenzweig, M. R. (1966). Environmental complexity, cerebral change, and behavior. *Amer. Psychol.* 21, 321–332.

Rosenzweig, M. R., Love, W., and Bennett, E. L. (1968). Effects of a few hours a day of enriched experience on brain chemistry and brain weights. *Physiol. Behav.* 3, 819–825.

Rowe, F. A., and Edwards, D. A. (1971). Olfactory bulb removal: Influences on the aggressive behaviors of male mice. *Physiol. Behav.* 7, 889–892.

Schreiner, L., and Kling, A. (1956). Rhinencephalon and behavior. *Amer. J. Physiol.* 184, 486–490.

Scott, J. P. (1958). *Aggression*, The University of Chicago Press, Chicago, Ill.

Segal, D. S., Knapp, S., Kuczenski, R. T., and Mandell, A. J. (1973). The effects of environmental isolation on behavior and regional rat brain tyrosine hydroxylase and tryptophan hydroxylase activities. *Behav. Biol.* 8, 47–53.

Seward, J. P. (1946). Aggressive behavior in the rat. IV. Submission as determined by conditioning, extinction and disease. *J. Comp. Psychol.* 39, 51–75.

Sheard, M. H. (1969). The effect of *p*-chlorophenylalanine on behavior in rats: Relation to brain serotonin and 5-hydroxyindoleacetic acid. *Brain Res.* 15, 524–528.

Smith, O. A. (1956). Stimulation of lateral and medial hypothalamus and food intake in the rat. *Anat. Rec.* 124, 363–364.

Uyeno, E. T., and Benson, W. M. (1965). Effects of lysergic acid diethylamide and attack behavior of male albino mice. *Psychopharmacologia* 7, 20–26.

Valzelli, L. (1966). Biological and pharmacological aspects of aggressiveness in mice. In: *Neuropsychopharmacology, Proc. C.I.N.P. Congress, Washington, D. C.*, March, 1966 (H. Brill, J. O. Cole, P. Deniker, H. Hippius, and P. B. Bradley, eds.), pp. 781–788. Excerpta Medica Foundation, Amsterdam, 1967.

Valzelli, L. (1967). Drugs and aggressiveness. *Adv. Pharmacol.* 5, 79–108.

Valzelli, L. (1971). Agressivité chez le rat et al souris: Aspects comportementaux et biochemiques. *Actual. Pharmacol. (Paris)* 24, 133–152.

Valzelli, L. (1973a). The "isolation syndrome" in mice. *Psychopharmacologia* 31, 305–320.

Valzelli, L. (1973b). Environmental influences upon neurometabolic processes in learning and memory. In: *Current Biochemical Approaches to Learning and Memory* (W. B. Essman, and S. Nakajima, eds.), pp. 29–47, Spectrum, New York.

Valzelli, L. (1974). 5-Hydroxytryptamine in aggressiveness. *Adv. Biochem. Psychopharmacol.* 11, 255–263.

Valzelli, L. (1976). Social experience as a determinant of normal behavior and drug effect. In: *Handbook of Psychopharmacology* (L. L. Iversen, S. D. Iversen, and S. H. Snyder, eds.), Plenum Press, New York, *in press*.

Valzelli, L., and Garattini, S. (1968). Behavioral changes and 5-hydroxytryptamine turnover in animals. *Adv. Pharmacol.* 6B, 249–260.

Valzelli, L., and Garattini, S. (1972). Biochemical and behavioral changes induced by isolation in rats. *Neuropharmacology* 11, 17–22.

Valzelli, L., Giacalone, E., and Garattini, S. (1967). Pharmacological control of aggressive behavior in mice. *Eur. J. Pharmacol.* 2, 144–146.

Vergnes, M., and Karli, P. (1964). Etude des voies nerveuses de l'influence facilitatrice exercée par les noyaux amygdaliens sur le comportement d'agression interspécifique rat-souris. *C. R. Seanc. Soc. Biol.* 158, 856–858.

Vergnes, M., and Karli, P. (1965). Etude des voies nerveuses d'une influence inhibitrice s'exerçant sur l'agressivité interspécifique du rat. *C. R. Seanc. Soc. Biol.* 159, 972–975.

Vergnes, M., and Karli, P. (1968). Activité électrique de l'ihippocampe et comportement d'agression interspécifique rat-souris. *C. R. Seanc. Soc. Biol.* 162, 555–558.

Vergnes, M., Mack, G., and Kempf, E. (1974). Contrôle inhibiteur du comportement d'agression interspécifique du rat: Système sérotoninergique du raphé et afférences olfactives. *Brain Res.* 70, 481–491.

Wasman, M., and Flynn, J. P. (1962). Directed attack elicited from hypothalamus. *Arch. Neurol.* 6, 220–227.

Welch, A. S., and Welch, B. L. (1968). Effect of stress and parachlorophenylalanine upon brain serotonin, 5-hydroxyindoleacetic acid and catecholamines in grouped and isolated mice. *Biochem. Pharmacol.* 17, 699–708.

Welch, A. S., and Welch, B. L. (1971). Isolation, reactivity and aggression: Evidence for an involvement of brain catecholamines and serotonin. In: *The Physiology of Aggression and Defeat* (B. E. Eleftheriou and J. P. Scott, eds.), pp. 91–142, Plenum Press, New York.

Welch, B. L. (1965). Psychophysiological response to the mean level of environmental stimulation: A theory of environmental integration. In: *Symposium of Medical Aspects of Stress in the Military Climate* (D. Mck. Rioch, ed.), pp. 39–96, Government Printing Office, Washington, D. C.

Welch, B. L., and Welch, A. S. (1969). Aggression and the biogenic amine neurohumors. In: *Aggressive Behaviour* (S. Garattini, and E. B. Sigg, eds.), pp. 188–202, Excerpta Medica, Amsterdam.

Wheatley, M. D. (1944). The hypothalamus and affective behavior in cats. *Arch. Neurol. Psychiat.* 52, 296–316.

Wood, C. D. (1958). Behavioral changes following discrete lesions of temporal lobe structures. *Neurology* 8, (Suppl. 1), 215–220.

Woods, J. W. (1956). "Taming" of the wild Norway rat by rhinencephalic lesions. *Nature (London)* 178, 869.

Yen, C. Y., Stanger, R. L., and Millman, N. (1959). Ataractic suppression of isolation-induced aggressive behavior. *Arch. Int. Pharmacodyn.* 123, 179–185.

Zeman, W., and Innes, J. R. M. (1963). *Craigie's Neuroanatomy of the Rat*, Academic Press, New York.

Behavioral Neurochemistry

7

Relationships among Rhythmic Slow Waves in the Brainstem, Monoamines, and Behavior[1]

TAKASHI TSUBOKAWA

Department of Neurological Surgery
School of Medicine
Nihon University
Tokyo, Japan

A. INTRODUCTION

It is known that an 8 to 10 per sec rhythm may be recorded in the ventral tegmental area (VTA), as described by Tsai (1926), and in the subthalamic nucleus (STN) and anterior raphé complex (Rh) under Nembutal anesthesia (Trembly and Sutin, 1962; Gahm and Sutin, 1968). This 8–10 per sec regular wave has been inhibited by lateral pallidal stimulation (Tsubokawa and Sutin, 1972), and disappears following a lesion in the anterior part of the Rh (Tsubokawa and Moriyasu, 1973). These regions may participate in the regulation of involuntary

[1]The author wishes to thank Dr. Jerome Sutin for the opportunity to work with him in the Department of Anatomy at Emory University, Georgia (U. S. A.). The first half of this experiment was done by Dr. Sutin and the author in his department. The author gratefully acknowledges the assistance of Dr. Takehito Sugawara and Dr. Toshikazu Goto with the biochemical measurements and fluorescence histochemistry in his department

movement (Carpenter and Strominger, 1967; Folkerts and Spiegel, 1953; Kaelber, 1963; Poirer, 1960) and in the sleep-waking cycle (Jalowiec et al., 1973; Morgane and Stern, 1972). Using histochemical fluorescence techniques, other investigators have suggested that monoaminergic neurons are located mainly in these regions (Dahlström and Fuxe, 1964; Falack et al., 1962) and that they play an important role in the regulation of involuntary movements and sleep.

To explain the generating mechanism of a rhythmic wave, relationships among electrophysiological phenomena, neurochemical findings, and behavioral alterations should be employed. For this purpose, the generating mechanism of a rhythmic wave and its physiological significance were investigated with electrophysiological and neurochemical methods, and the data were correlated with observations of animal behavior.

B.   MATERIALS AND METHODS

Adult cats weighing between 2 and 3.5 kg were used. For electrophysiological experiments, 46 animals were anesthetized by intraperitoneal injections of 25–30 mg/kg sodium pentobarbital (Nembutal). Following tracheotomy and cannulation of the femoral vein, animals were placed in a Horsley–Clarke apparatus and burr holes were made in their skulls for electrode placement. In order to study electrical activity in the absence of Nembutal or the effects of administration of monoamines and inhibitors of monoamine metabolism, 23 additional animals were anesthetized with ether for the duration of surgical preparation and then lidocaine (1%) was used to infiltrate all wound edges and pressure points. Animals were then immobilized with gallamine triethiodide and artificially respirated.

For stimulation and recording, bipolar stainless steel electrodes (26 gauge) were used. Stimulus pulses were usually of 0.5 msec duration, 0.1–0.5 peak mA. intensity, and 1 per sec frequency. Vinyl lacquer insulated tungsten wire or glass pipettes filled with 4 M sodium chloride were used for recording. Unitary activities and rhythmic waves were stored on FM magnetic tape for subsequent computer analysis. An RCA-Spectra 55 computer was used to construct interspike interval and poststimulus histograms. The auto and cross-correlation technique was based on the method introduced by Gogolak et al. (1968). After the experiments, animals were perfused with saline, followed by 10% formol-saline. Recording sites were identified in frozen brain sections stained by cresyl violet.

For neurochemical experiments, 10 animals were anesthetized with

Nembutal and 12 additional animals were initially anesthetized by intramuscular injection of ketamine (Ketaral; 5–10 mg/kg). Then, with local anesthesia, bipolar electrodes were inserted into the VTA, STN, and Rh to record rhythmic slow waves. Nembutal, norepinephrine, 5-hydroxytryptophan (5-HTP), and harmaline (an inhibitor of 5-HT oxidase) were administered by intravenous injection or by local micro-injection into the VTA or STN. After recording electrical activity in these regions, the brain was exposed and frozen with liquid nitrogen. Approximately 1 g of tissue was excised from both the VTA and STN. The contents of serotonin (5-HT), dopamine (DA), and norepinephrine (NE) in the homogenized brain tissue were determined using the spectrophotofluorometic method of Fleming et al. (1965).

For behavioral experiments, a metal cannula, which served as a guide for insertion of the injection needle, was implanted stereotactically through the burr hole and fastened with dental cement. Four to five days after the operation, a 28-gauge injection needle, with insulated stainless wire electrodes for recording the activities of the drug injected area, was inserted through the implanted metal cannula. 5-HTP (1 mg/ml), harmaline (3 mg/ml), and NE (3 mg/ml) were administered using a microinjector. The total volumes of injected substances were always less than 0.3 ml. At the end of each experiment, a volume of methylene blue equivalent to that of the drug used was injected into the same region for histological identification of the diffusion area. With the animals unrestrained on a stage, observations were made of sensorimotor behavior and sleep. Electrical activities in the free-moving animals were also recorded. In 5 cases, movements were recorded by time-lapse photography, taking one frame every 0.5 sec with an electric motor-driven Nikon camera.

C. RESULTS

1. Rhythmic Waves and Evoked Potentials between the STN and the VTA

Under Nembutal anesthesia rhythmic waves were recorded in the STN, VTA, and anterior part of mesencephalic raphé nucleus. Evoked potentials in the STN, produced by VTA stimulation, showed early negative waves with a 7–12 msec latency followed by an 80- to 120-msec positive phase which lasted for as long as the positive phase of rhythmic waves. Evoked potentials of similar characteristics, produced by STN stimulation, were recorded in the VTA.

Rhythmic waves following stimulation of the VTA, STN, or Rh

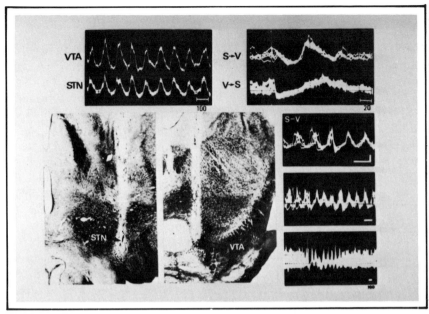

1 At the top left, the spontaneous rhythmic slow waves recorded in VTA and STN are shown. Recording sites are indicated by arrows (lower left). At the upper right, are shown evoked potentials in the VTA produced by STN stimulation (S→V) and those in the STN produced by VTA stimulation (V→S). The right side of all figures indicates the phase-locked phenomenon of evoked potentials and spontaneous slow waves in the STN following VTA stimulation (sweeps are superimposed).

constitute a phase-locked phenomenon, lasting for 2 sec (Fig. 1). These waves and evoked potentials in both the VTA and STN could not be recorded under local anesthesia or following intravenous administration of epinephrine, amphetamine, and dopamine. However, the rhythmic wave could be induced by intravenous injection of 5-HTP or harmaline in locally anesthetized preparations (Fig. 2). The minimum amounts necessary to induce the rhythmic slow wave were 9 mg/kg of 5-HTP and 30 mg/kg of harmaline during 5-HTP or harmaline administration. The shapes of evoked potentials between VTA and STN resembled those recorded in Nembutal-injected animals, except for a small, abortive, negative wave with 25-msec latency in the first positive phase.

2. Unit Activity of the VTA, the STN, and the Rh during Rhythmic Wave Recording

When the 8–12 per sec rhythmic wave was present in VTA, STN, and Rh, 90 STN units responded to VTA stimulation. In the VTA 101

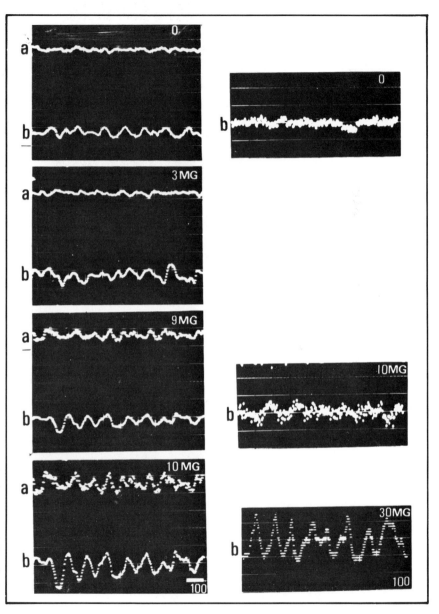

2 Rhythmic waves in the STN (a) and in the VTA (b) were induced following injection of 5-HTP (left side) and harmaline (right side). MG indicates the total amount of intravenously injected 5-HTP or harmaline.

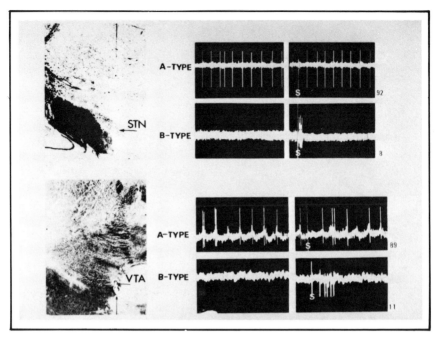

3 Unit activities of the STN (top) and the VTA (bottom). In the left side, recording sites are indicated. In the STN, spontaneous firing and both excilatory and inhibitory effects ("A-type" cell) and burst firing ("B-type" cell) by VTA stimulation are shown. In the VTA, the same types of cells were found, as seen at the bottom.

units responded to STN stimulation, and 76 Rh units responded to VTA or STN stimulation.

In response to VTA stimulation, 92% of STN units showed regular spontaneous firing and driven potentials with a 6.0 $\pm$ 2.0 msec latency, followed by a 80–120 msec inhibitory period. These were termed "A-type" cells. Another 8% of all units showed bursts with 16.2 msec latency, coinciding with the duration of the inhibitory phase of an "A-type" cell response to VTA stimulation. The spontaneous bursts were irregular and of low frequency. This type of neuron was termed "B-type" cell (Fig. 3).

In 89% of all VTA units, STN stimulation evoked spontaneous regular firing of "A-type" cells and driven action potentials with 11.0 $\pm$ 4.0 msec latency followed by a 100.5 $\pm$ 4.8 msec inhibitory period which corresponded to the inhibitory period of "A-type" cells (see Fig. 3). The remaining 11% of VTA units responded after 33.3 msec of latency and also showed irregular, low-frequency spontaneous firing.

The interspike interval histogram of spontaneously firing A-type

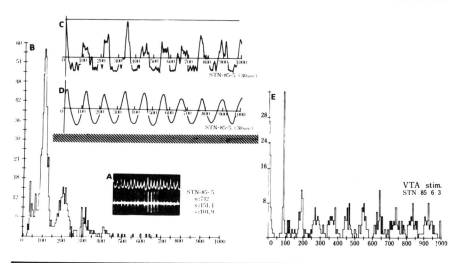

4 A. Spontaneous firing of an "A-type" cell and rhythmic slow wave recorded in the STN. B. Interspike interval histogram. C. Autocorrelogram of the spike. D. cross-correlogram of the spike and slow wave. E. Poststimulus histogram of the firing of an "A-type" cell of the STN produced by VTA stimulation.

cells had two patterns. One closely approximated a Gaussian distribution with 96.5-msec mean interval; while the other had several peaks separated by approximately 100 msec in its distribution curve (Fig. 4B). The possibility that regular unit discharges coincided with the rhythmic wave was assessed by examination of the temporal correlation between spikes (autocorrelograms). The top correlogram in Figure 4C shows the spike train correlated with itself. The cross correlogram shown in Figure 4D indicates the high degree of correlation between unit activities and negative peaks of rhythmic waves.

In the poststimulus histogram of STN neurons under VTA stimulation, the higher firing probability is correlated with both the negative peaks of the evoked potentials and the phase-locked rhythmic wave, as was the case with evoked potentials between VTA and STN (Fig. 4E). The STN histogram revealed characteristics similar to those of the VTA "A-type" cells.

In the Rh, 76 units had a regular firing "A-type" like those of the VTA or the STN, but no "B-type" cells were found. These "A-type" cells were located in the anterior raphé complex (i.e., nucleus raphé

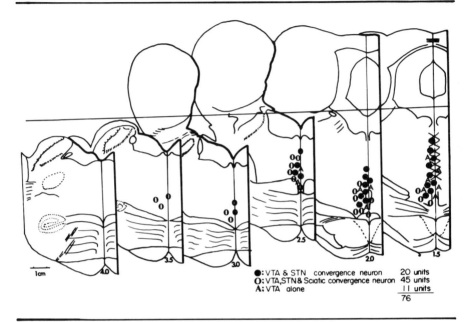

●: VTA & STN convergence neuron	20 units	
O: VTA,STN & Sciatic convergence neuron	45 units	
A: VTA alone	11 units	
	76	

5 The distribution of Rh neurons responding to STN and VTA stimulation. Numbers at the bottom indicate the distance from the 0 point of a stereotaxic map of the cat brain.

dorsalis, nucleus linearis rostalis, and nucleus raphé medialis; see Fig. 5). This area corresponds to B_9, B_8, and B_7 of the serotoninergic neuron, according to Dahlström and Fuxe (1964). The "A-type" cell of the Rh showed a 104.0 ± 36.3 msec inhibitory phase after VTA stimulation, and a 96.0 ± 25.3 msec inhibitory phase after STN stimulation, without constant driven activity (Fig. 6). Of the Rh "A-type" cells, 26% exhibited convergent input from both VTA and STN stimulation, and 59% of the "A-type" cells had convergent input from STN, VTA, and sciatic stimulations. The inhibitory period produced by sciatic stimulation was longer than that produced by VTA or STN stimulation. These neurons were located mainly in the ventral, lateral, and caudal marginal regions of the rostral raphé nucleus (Fig. 5). Another 4% of the "A-type" cells responded only to VTA stimulation.

By stimulation of the region of Rh containing "A-type" cells, short latency-driven action potentials followed by 80- to 100-msec inhibitory periods could be recorded in VTA and STN neurons (Fig. 6). The driven action potential latency was 6.0–7.0 msec for these VTA "A-type" cells and 8.5–9.0 msec for those of the STN.

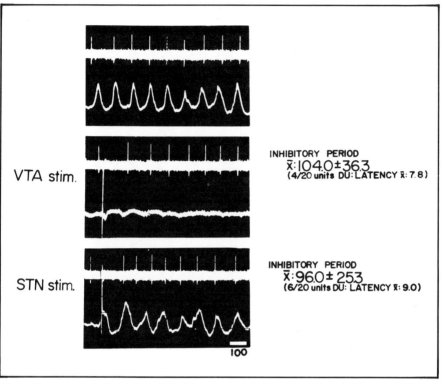

6

Unit activity of an Rh neurone that responded to VTA and STN stimulation. Top. Spontaneous firing with VTA rhythmic wave. Center. Response to VTA stimulation with STN evoked potential and rhythmic wave. Bottom. Response to STN stimulation with VTA evoked potential and rhythmic wave.

The poststimulus histogram of action potentials of both "A-type" cells of STN and VTA neurons elicited by Rh stimulation shows a higher firing probability (see Fig. 7), corresponding well with the negative evoked potential peaks and phase-locked rhythmic wave cause by Rh stimulation. After lesioning the anterior raphé complex (Fig. 8), the rhythmic wave disappeared and the "A-type" cells of both VTA and STN fired irregularly. The positive phase of evoked potentials between VTA and STN also disappeared. As shown on the right side of Fig. 8, the VTA "A-type" cells responded to STN stimulation only with an early, driven action potential.

Analysis of these electrophysiological findings revealed that a reciprocal fiber connection exists between "A-type" cells of the VTA and STN and that axon collaterals of these neurons are connected with

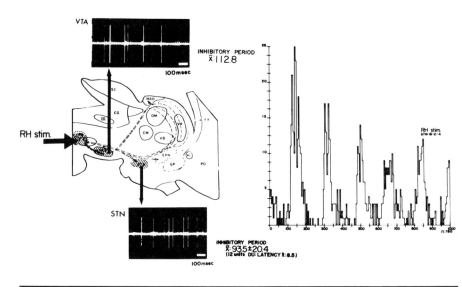

7   Action potentials in "A-type" cells of the VTA (VTA) and the STN (STN) following Rh stimulation. A poststimulus histogram of "A-type" cells of the STN is shown on the right side.

"B-type" cells whose axons connect with "A-type" cells. "B-type" cells exert inhibitory effects on "A-type" cells, and also receive impulses from the Rh. This anatomic circuitry forms the basis of production of localized rhythmic waves in the mesencephalon. Therefore, rhythmic waves can be induced by depolarization and hyperpolarization of the synchronized membrane potentials of "A-type" cells of the VTA and STN. Also in relation to this anatomic circuitry, Nembutal, serotonin, and harmaline increased the duration of bursts to 80–120 msec. These findings suggest that the anterior raphé performs an important role by lengthening the period of bursts of "B-type" cells. In both VTA and STN "B-type" cells are fired by Rh stimulation, and, moreover, the duration of their bursts is increased in proportion to production of rhythmic waves when serotonin or a serotonin oxidase inhibitor are applied.

3.   Monoamine Content of the STN and the VTA in the Presence of Rhythmic Waves Induced by Administration of Nembutal, 5-HTP, and Harmaline

While rhythmic waves were recorded in animals under Nembutal anesthesia or injected intravenously with 5-HTP and harmaline, the

Table I. The Alteration in the Contents of 5-HT, DA and NE in the VTA and the STN by Intravenous Injection of Nembutal (Surgical Doses), 5-HTP (10mg) and Harmaline (30mg). In All Cases Rhythmic Slow Waves Were Evident in Both VTA and STN.

	V T A				S T N			
	Normal	Nembutal	5HTP	Harmaline	Normal	Nembutal	5HTP	Harmaline
5 H T	0.5	6.2	2.5	1.4	0.8	7.1	2.4	1.3
D A	0.2	5.1	0.5	0.4	0.3	5.3	0.4	0.4
N E	2.3	5.3	1.8	2.6	3.4	5.7	1.6	2.4

experiment was terminated by freezing the brains with liquid nitrogen. About 0.5–1.0 g of brain tissue was excised from both the VTA and STN for determination of dopamine, norepinephrine, and serotonin content(s). The levels of these monoamines were compared with those obtained from locally anesthetized cats that did not display rhythmic waves in both the VTA and STN. The monoamine contents of the VTA of control animals were 0.5 μg/g of serotonin, 0.2 μg/g of dopamine, and 2.3 μg/g of norepinephrine. Their STN regions contained 0.8 μg/g of serotonin, 0.3 μg/g of dopamine, and 3.4 μg/g of norepinephrine. In the presence of rhythmic waves recorded under Nembutal anesthesia, both serotonin and dopamine contents rose to 8–10 times control levels, i.e., to about 5.0–6.0 μg/g. Norepinephrine content was increased to about 1.5–2.0 times the control level (Fig. 9). These results indicated that a monoaminergic neuronal factor may play a significant role in lengthening the duration of bursts of "B-type" cells.

The serotonin contents of the VTA and STN rose to 3–5 times the control level when rhythmic waves were induced by intravenously injected 5-HTP (9–10 mg/kg), whereas dopamine was not markedly increased and norepinephrine was decreased to 78% of its control level in the VTA and to 47% of its control level in the STN (Fig. 9). In harmaline-treated animals, the results were similar to those obtained in 5-HTP-treated animals (Fig. 9).

These biochemical findings suggest that the firing period of the "B-type" cell is lengthened by an increase of serotonin content in both the VTA and STN. In locally anesthetized animals, fewer cells contain 5-HT in both the STN and VTA (Dahlström and Fuxe, 1964) supporting our biochemical data. In order to elucidate further the relationship between the alteration of serotonin cell content and the appearance of rhythmic waves produced by administration of Nembutal, 5-HTP, and harmaline, fluorescence histochemical studies were performed in the VTA and STN. In the serotonin- or harmaline-treated animals which

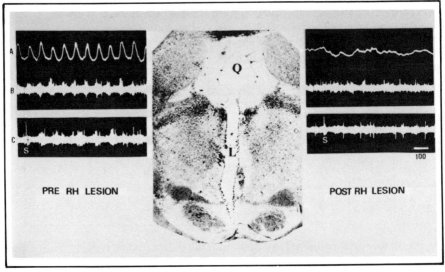

8 Experimental modifications of the VTA rhythmic wave.
A. Spontaneous firing of multiple units of "A-type" cells of the VTA. B. Action potentials of the VTA that responded to STN stimulation. C. Rhythm wave activity following a lesion of the rostal raphe nulcei (L).

exhibited rhythmic waves in both the VTA and STN, the large cells of the STN, presumed to be "A-type" cells, were stained a characteristic greenish color, indicating that they were dopaminergic cells although their reaction with thionyl chloride was not tested. These dopaminergic cells had several fine networks of serotonergic terminals around their surfaces identified by the yellowish fluorescence.

Intermixed with "A-type" cells, there were other small cells, presumed to be "B-type," which exhibited a typical yellowish fluorescence indicating that they might be serotoninergic (see Fig. 10). Some axons of these "B-type" cells also had a yellowish fluorescence. In the VTA, serotoninergic small cells were more conspicuous than in the STN (Fig. 11). After lesioning of the Rh, rhythmic waves disappeared, and serotoninergic small cells and fine terminals on dopaminergic cells could no longer be found in VTA or STN.

4. Relationships between Increase of Serotonin in VTA and STN and Behavioral Changes

In chronically cannulated animals, 0.1 mg of 5-HTP or 0.3 mg of harmaline were introduced into the left VTA by means of a micro-injector. The animals became quiet, easy to handle, and did not vocalize. Rhythmic waves were recorded at the site of drug injection. Muscle

144

tone was normal; no muscular rigidity or ataxic gait was noted. Animals performed contralateral circling movements whenever they tried to walk or run, which continued until they collapsed into curled positions (Fig. 12). The animals could walk erectly, however, if the right lateral sides of their bodies were in contact with a straight wall, or if the right side of the head or the body surface was touched. Rhythmic waves were also inhibited by tactile stimulation of the contralateral side of the body near the area of 5-HTP injection (Fig. 13). Although rhythmic waves could be recorded in the VTA or STN after administration of Nembutal in acute experiments, these waves were not altered by tactile stimulation of the body surface. However, "A-type" cells of the Rh neuron were inhibited for more than 120 msec by sciatic stimulation, even after Nembutal anesthesia.

Microinjections of 5-HTP and harmaline into the red nucleus or the midbrain reticular formation did not induce any behavioral alterations. However, the spread of 5-HTP or harmaline to the substantia nigra, especially its mediodorsal aspect, caused by excess injection volume, induced contralateral circling movements. This circling could not

9 Alteration in 5-HT, DA, and NE contents in the VTA and STN following intravenous injection of Nembutal (surgical doses), 5-HTP (10 mg), and harmaline (30 mg). In all cases, rhythmic slow waves were evident in both VTA and STN.

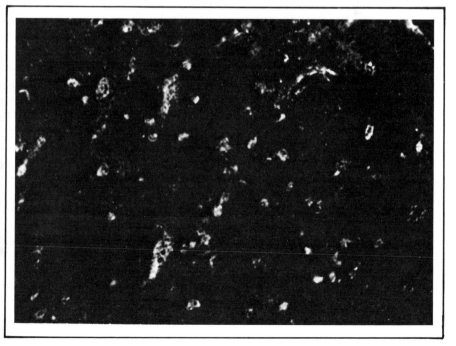

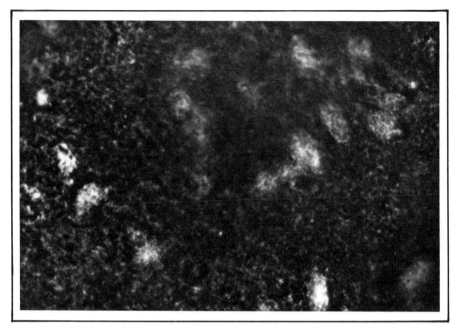

10   Fluorescence histochemical findings in the STN following treatment with 5-HTP.

be arrested by tactile stimulation on the contralateral body surface or by body contact with the wall. These experiments indicated that the following factors play significant roles in the production of localized rhythmic waves: (a) An anatomically specific fiber connection exists between the VTA and STN, consisting of dopaminergic neurons, presumed to be "A-type" cells. Axon collaterals of "A-type" cells are connected to "B-type" cells which are serotoninergic neurons, and serotonergic neurons are also connected to "A-type" cells. (b) The firing period of serotoninergic cells, presumed to be of the "B-type," is increased by about 80–120 msec when the serotonin content of "B-type" cells rises. Norepinephrine or dopamine contents in "A-type" cells bear no relation to the production of rhythmic waves. Nembutal might inhibit the action of monoamine oxidase as does harmaline, and thereby produce an increase in serotonin in the serotoninergic cells. (c) The anterior raphé complex connects with "B-type" (serotoninergic) cells; augmented Rh activity induces a rise in the serotonin content of these serotoninergic cells and in their axon terminals which impinge on dopaminergic cells. (d) Rhythmic waves are intimately related to some kinds of involuntary movements, modulated by peripheral proprioceptive inputs.

146

D. DISCUSSION

1. Generating Mechanism of Rhythmic Waves in the Mesencephalon in Relation to Monoamines

Results presented here indicate that reciprocal fiber connections between the main VTA and STN cells, whose axon collaterals connect with burst cells receiving impulses from the Rh, are prerequisite for production of localized rhythmic waves. In order to generate rhythmic waves in these specific anatomic structures, the duration of bursts of "B-type" cells which inhibits "A-type" cells must be prolonged by administration of Nembutal, 5-HTP, or harmaline.

Eccles et al. (1963) found that Nembutal and chloralose increased the time course of spinal cord presynaptic inhibition, and the recurrent IPSP's of motorneurons have also been prolonged by barbital (Larson and Major, 1970) without any concomitant change of duration of Renshaw cell discharge. These investigators suggested that this effect could be due to a restriction of diffusion of the transmitter from the synaptic cleft. In our experiments, however, the duration of bursts of "B-type" cells was augmented by increasing the concentrations of sero-

11 Fluorescence histochemical findings in the VTA following treatment with 5-HTP.

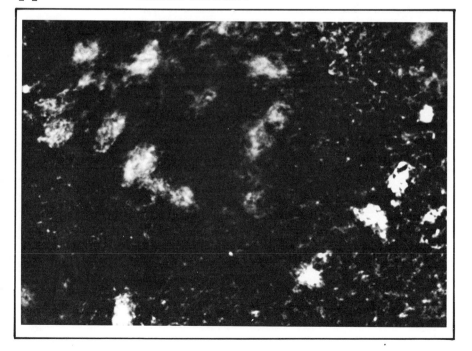

tonin, dopamine, and norepinephrine in the VTA and STN. Therefore, the prolonged inhibitory period did not appear to be caused by restriction of transmitters in the synapses, but rather by serotonin itself.

Biochemical results indicated that Nembutal might act as a monoamine oxidase inhibitor in the VTA, STN, and Rh in the vicinities of monoaminergic neurons, since serotonin (5-HT), dopamine, and norepinephrine were increased in these areas, and since 5-HT was increased both in the interneurons presumed to be "B-type" cells and in axon terminals, as shown by fluorescence histochemistry. Furthermore, administration of 5-HTP under local anesthesia produced rhythmic waves and prolonged the duration of bursts of "B-type" cells, and harmaline (a 5-HT oxidase inhibitor) had the same effect as injected 5-HTP.

Although Montigny and Lamarre (1973) also induced rhythmic activity by harmaline injection into the olivocerebellobulbar system and recorded bursts from neurons of the fastigal, the bulbar reticular, and the lateral vestibular nuclei, they could not explain the mechanism by which harmaline induces rhythmic firing of olivary neurons. As shown by both biochemical determinations and fluorescence histochemical re-

12 Circling movements were recorded every 0.5 sec by time-lapse photography following local microinjection of 5-HTP. The histology shows the area of spread of injected methylene blue. Spontaneous rhythmic waves were recorded at VTA.

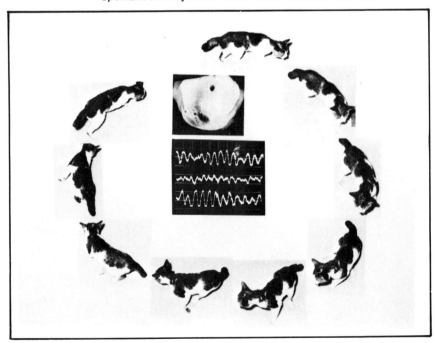

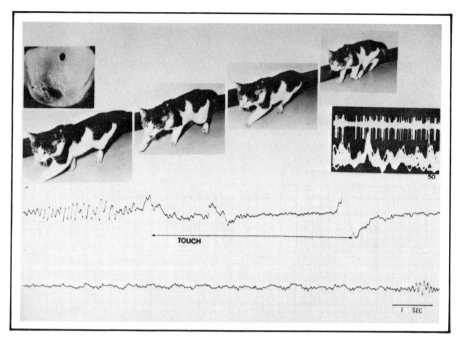

13

This cat, injected with 5-HTP in the VTA, can walk straight whenever the contralateral side of the body surface makes contact with the wall or with something else. The rhythmic wave is suppressed during and after tactile stimulation.

sults, increases in the serotonin contents of small interneurons ("B-type" or serotoninergic cells) and their axon terminals which connect with "A-type" or dopaminergic cells are prerequisite for the production of rhythmicity. The finding that stimulation of the raphé nucleus, which contains the major portion of serotonergic cells, increased the rhythmicity of both the VTA and STN supports the theory that a connection exists between serotonin and "B-type" cells.

It might be concluded that the increased duration of firing of "B-type" cells produces hyperpolarization of membrane potential of the "A-type" cells which is reflected as the positive phase of rhythmic waves. This prolongation may be caused by an increase in the serotonin content of "B-type" cells that can be induced by direct injection of 5-HT or of 5-HT oxidase inhibitors like harmaline or Nembutal.

2. Relationship between the Generation of Rhythmic Waves and Behavioral Alterations

Rhythmic 8–12 per sec waves could be recorded in both the STN and VTA in the presence of levels of Nembutal anesthesia which produced barbiturate tremor in acute experiments. No tremors were observed, however, in chronic animals injected with 5-HTP or harmaline, even though rhythmic waves could be recorded in the VTA and STN. Therefore, the neuronal mechanism involved in barbiturate-induced tremor does not appear to be related to the neural mechanism involved in the generation of rhythmic waves. Tremor could not be induced by administration of serotonin, but harmaline enhanced postural tremor following ventral tegmental lesions (Poirier et al., 1966). Since harmaline prevents the degradation of serotonin, Poirier et al. (1966) suggested that serotonin might be affected by dopamine metabolism in the striatum. In our experiments, however, tegmental area lesions produced no increase of 5-HT or harmaline.

Following microinjection of 5-HTP or harmaline into the VTA in chronic, awake animals, unexplained contralateral circling movements were observed while recording rhythmic waves in the VTA. Contralateral circling movements induced by microinjection of serotonin or harmaline in the absence of changes in muscle tone and rigidty have not been previously described, although many monoaminergic fiber connections may serve to control the extrapyramidal motor system in the VTA and STN.

Myers (1964) reported that injection of serotonin or 5-HTP into the substantia nigra induced tremor or circling movements 15–100 min after injection and lasting for 30–150 min. Circling movements have also been produced by application of cholinergic substances to the caudate nucleus, substantia nigra, red nucleus (Nashold and Gills, 1960), and lateroposterior parts of the hypothalamus (Myers, 1964). In our experiments, injection of 5-HTP or harmaline into the substantia nigra evoked very pronounced circling movements that could not be stopped by the tactile stimulation of the contralateral body surface, which was effective in blocking circling movements that followed injection into the VTA. This finding does not mean that injected serotonin spreads into the substantia nigra, red nucleus, and caudal midbrain. With Nile blue and radioautographic techniques, MacLean (1957) demonstrated that powdered chemical agents injected into brain tissue spread little more than a 1 mm radius in 40 min, and biogenic amines (e.g., dopamine 1 μg in 1 μl saline) spread rostrally and caudally for respective total distances of 2.1 and 1.8 mm, via extracellular channels (Bordareff et al., 1970).

Some evidence indicates that our injected substances remained localized because no circling movements were induced by injection of 5-HTP or harmaline into the red nucleus, and opticokinetic nystagmus was observed just after the circling movement, indicating that spread of injected serotonin did not invade the caudal midbrain. Since circling movements have been induced by tonic electrical stimulation, by carbachol injection, or even by making lesions in the substantia nigra or red nucleus (Nashold and Gills, 1960), the induction of circling movements may not be directly related with these latter nuclei. Therefore, circling movements may have been induced by increased serotonin in the VTA and in other regions which have serotoninergic neuronal connections with the VTA. There is no direct evidence that rhythmic waves are generated by, or associated with the appearance of circling movements, even though the increase of serotonin in the VTA induces circling movements and localized rhythmic waves.

3. The Relationship between Rhythmic Waves and Sleeping

Raphé lesions abolished rhythmic waves in the VTA and STN, and stimulation of the Rh induced phase-locking of rhythmic slow waves in VTA and STN in these experiments. After intravenous injection of serotonin (Koella and Czicman, 1966), or after its injection into the bulbopontine region (Ledebur and Tissot, 1966), ocular and EEG signs of synchronized EEG sleep are quickly seen. Morgane and Stern (1972) have recently summarized results indicating that injection of serotonin into the midbrain raphé nuclei of the cat does not affect the animal's sleep profile, but that depletion of central serotonin stores increases total sleep time without affecting paradoxical sleep. The purpose of the present investigation was not to study the "sleep-wakefulness" pattern, but we observed that with increased VTA serotonin, the cats were easier to handle. The clear-cut correlation suggested by Jalowiec et al. (1973) between regional levels of serotonin in the VTA and sleep-waking profiles has not yet been established.

E. SUMMARY

The generating mechanism and behavioral significance of the mesencephalic rhythmic wave were studied by unitary discharge recording; chemical analysis of local norepinephrine, dopamine, and serotonin content using Fleming's method; distribution of monoamine neurons with Falck and Hillarp's method; and microinjection of monoamines and monoamine oxidase inhibitors.

1. In the VTA, STN, and Rh where rhythmic waves were recorded, there are two types of cells; the main cell type shows regular firing related with the negative peak of rhythmic wave, and the other cell

type exhibits bursts corresponding with the positive phase of rhythmic waves. There are none of these firing cells in the Rh. Analysis of single unitary recordings revealed a reciprocal connection between the VTA and STN, with postsynaptic inhibitory synapses receiving impulses from the Rh. This anatomic circuitry may be involved in the production of localized rhythmic waves.

2. The monoamine contents of the VTA and STN, during appearance of rhythmic regular waves in cats under Nembutal anesthesia, differ from those of normal cats. Serotonin increased to 5–8 times and norepinephrine increased to 1.5–2.0 times control values.

3. Rhythmic regular waves in the VTA and STN can be induced by intravenous injection of 5-HTP (9–10 mg/kg) or harmaline (20–30 mg/kg), but not by Nembutal injection. In both the VTA and STN 5-HT increased 4–5 times above its normal value, but norepinephrine in the STN decreased 78%. Similar alterations of monoamines in the VTA and STN were observed when rhythmic regular waves were induced by injection of harmaline. These findings suggest that rhythmic regular waves were generated in the neural circuit between the VTA and STN which could be triggered by Rh activity when the firing time of bursts was extended to 80–120 msec by accumulation of serotonin.

4. The monoamine synaptic connections in the VTA, STN, and Rh were studied by Falck and Hillarp's histochemical method. The main cells of the VTA and STN did not contain serotonin in their cell bodies and were dopaminergic, but serotonin was observed around the main cells and in cells with burst activity when rhythmic waves were recorded in both the VTA and STN.

5. Microinjections of Nembutal, 5-HTP, or harmaline into the VTA of the cat induced contralateral circling movements which were blocked by tactile stimulation of the contralateral side of the body.

F. REFERENCES

Bordareff, W., Routtenberg, A., Narotzky, P., and McLone, D. G. (1970). Intrastriatal spreading of biogenic amines. *Exp. Neurol.* 28, 213–229.

Carpenter, M. D., and Strominger, N. C. (1967). Efferent fibers of the subthalamic nucleus in monkey. *Amer. J. Anat.* 121, 41–72.

Dahlström, A., and Fuxe, K. (1964). Evidence for the existence of monoamine containing neurones in the central nervous system. *Acta. Physiol. Scand.* 62, suppl. 232.

Eccles, J. C., Schmidt, R., and Willis, W. D. (1963). Pharmacological studies on presynaptic inhibition. *J. Physiol. (London)* 168, 500–530.

Falack, B., Hillarp, N. A., Thieme, G., and Torp, A. (1962). Fluorescence of catecholamines and related compounds condensed with formaldehyde. *J. Histochem. Cytochem.* 10, 348–354.

Fleming, R. M., Clark, W. G., Fenster, E. D., and Towne, J. C. (1965). Single extraction method for simultaneous fluorometric determination of serotonin, dopamine and norepinephrine in brain. *Anal. Chem.* 37, 692–696.

Folkerts, J. F., and Spiegel, E. A. (1953). Tremor on stimulation of the midbrain tegmentum. *Confin. Neurol.* 13, 193–202.

Gahm, N. H., and Sutin, J. (1968). The relation of the subthalamic and habenular nuclei to oscillating slow wave activity in the midbrain ventral tegmental area. *Brain Res.* 11, 507–521.

Gogolak, G., Stumpf, C. H., Petsche, H., and Sterc, J. (1968). The firing pattern of septal neurones and the form of the hippocampal theta wave. *Brain Res.* 7, 201–207.

Goldstein, M., Battista, A. F., Nakatani, S., and Anagoste, B. (1969). The effect of centrally acting drugs on tremor in monkey with mesencephalic lesion. *Proc. Nat. Acad. Sci. U.S.A.* 63, 1113–1116.

Jalowiec, J. E., Morgane, P. J., Stern, W. C., Zolovick, A. J., and Panksepp, J. (1973). Effect of midbrain tegmental lesions on sleep and regional brain serotonin and norepinephrine levels in cats. *Exp. Neurol.* 41, 670–682.

Kaelber, W. W. (1963). Tremor at rest from tegmental lesions in the cat. *J. Neuropathol. Exp. Neurol.* 22, 695–701.

Koella, W. P., and Czicman, J. (1966). Mechanisms of the EEG-synchronizing action of serotonin. *Amer. J. Physiol.* 211, 926–934.

Larson, M. D., and Major, M. A. (1970). The effect of hexobarbital on the duration of recurrent IPSP in cat motoneurones. *Brain Res.* 21, 309–311.

Ledebur, I. X., and Tissot, R. (1966). Modification de l'activité électrique cérébrale du lapin sous l'éffect de micro-injections de précurseurs des monoamines dans les structures somnogénes bulbaires et pontiques. *EEG Clin. Neurophysiol.* 20, 370–381.

MacLean, P. D. (1957). Chemical and electrical stimulation of hippocampus in unrestrained animals. I. Methods and electroencephalographic findings. *A.M.A. Arch. Neurol. Psychiat.* 78, 113–127.

Montigny, C. De, and Lamarre, Y. (1973). Rhythmic activity induced by harmaline in the olivo-cerebello-bulbar system of the cat. *Brain Res.* 53, 81–95.

Morgane, P. J., and Stern, W. C. (1972). Relationship of sleep to neuroanatomical circuits, biochemistry, and behavior. *Ann. N. Y. Acad. Sci. U.S.A.* 193, 95–111.

Myers, R. D. (1964). Emotional and autonomic responses following hypothalamic chemical stimulation. *Canad. J. Psychol.* 18, 6–14.

Nashold, B. S., and Gills, J. D., (1960). Chemical stimulation of telencephalon, diencephalon, and mesencephalon in unrestrained animals. *J. Neurol. Path. Exp. Neurol.* 19, 580–590.

Poirier, L. J. (1960). Experimental and histological study of midbrain dyskinesias. *J. Neurophysiol.* 23, 534–551.

Poirier, L. J., Sourkes, T. L., Bouvier, G., Boucher, R., and Caravin, S. (1966). Striatal amines, experimental tremor and the effect of harmaline in the monkey. *Brain* 89, 37–52.

Poirier, L. J., McGeer, E. G., Larochelle, L., McGeer, P. L., Bedard, P., and Bouchor, R. (1969). The effect of brain stem lesions on tyrosine tryptophan hydroxylase in various structures of the telencephalon of the cat. *Brain Res.* 14, 147–155.

Tsai, C. (1926). The optic tracts and centers of opposum, *Didelphis virginiana. J. Comp. Neurol.* 39, 173–216.

Trembly, B., and Sutin, J. (1962). Slow wave activity in the ventral tegmental area of Tsai related to barbiturate anesthesia. *Exp. Neurol.* 5, 120–130.

Tsubokawa, T., and Sutin, J. (1972). Pallidal and tegmental inhibition of oscillatory slow wave and unit activity in the subthalamic nucleus. *Brain Res.* 41, 101–118.

Tsubokawa, T., and Moriyasu, N. (1973). The generating mechanism of rhythmic slow wave in the ventral tegmental area. *Clin. EEG* (Jap. ed.) 15, 556–564.

8

Influence of Some Mediators upon the Metabolic Activity of Neuronal Subcellular Structures[1]

NINA DOCHEVA GEORGIEVA
RADOY IVANOV IVANOV

Department of Human and Animal Physiology
Faculty of Biology
University of Sofia
Sofia, Bulgaria

A. INTRODUCTION

The main functions of neurons are the generation and spread of nervous impulses and the integration of excitement and retention processes (e.g., memorizing). These functions are closely linked to energetic and plastic neuronal functions. Mediators and neurohormones regulate neural activity by controlling changes in membrane permeability, in the courses and characters of the enzyme-chemical processes, and in the activities of various cytoplasmic and subcellular structures. To a great extent, the metabolic and functional states of cells are regulated by processes in mitochondria, which are among the most highly reactive structures of the cell.

Changes in the activities and concentrations of mediator systems

[1]The authors are grateful to Prof. L. A. Kalchev for advice and to Vessela Georgieva and Elgina Steffanova for their assistance in preparing this manuscript.

and biologically active substances in microstructures at various micro-intervals are perhaps involved in the modulation of cellular metabolism and function. Many workers have looked for cause and effect links between redox reactions and protein biosynthesis (ref. 16), functional state (Barreth and Beis, 1973; Ivanov et al., 1974; see ref. 18), generation of bioelectric impulses, pharmaceuticophysiological sensitivity, and reactivity (Dutton and Wilson, 1974). The research and theoretical treatments of Szent Györgyi (see ref. 14) and Pullman (1965) have provided a rather promising perspective on these relationships. Difficulties exist in defining the redox potentials of heterogeneous biological systems, and in finding adequate apparatus for performing microanalyses on highly differentiated abilities *in vivo* (Dutton and Wilson, 1974). This report discusses some of our investigations on the redox state of brain mitochondrial suspensions and the effects of acetylcholine (ACh), γ-aminobutyric acid (GABA), and epinephrine (Adr) on this phenomenon. The actions of these agents were studied in a pure physical–chemical system.

B. MATERIALS AND METHODS

Cat brain mitochondria were isolated according to the method of Lovtrup and Zelander (1962). Redox potentials were studied in thermostatic glass cuvettes, the reactants being electromechanically mixed at the time of incubation. Platinum and calomel electrodes were used, in some cases the latter being joined to the system by means of an electrolyte switch in order to eliminate diffusion potentials. Changes in redox state were recorded after passing signals through a preamplifier.

In some cases, electrochemical potential differences were registered by means of chlor-silver electrodes. Mitochondrial metabolic states *in vitro* were modulated by varying some components of the incubation media. The media used were similar to those used by Warburg in his classic experiments on oxygen consumption and oxidative phosphorylation.

C. RESULTS AND DISCUSSION

Figure 2 shows a change in electrical charge between two cuvettes containing incubation medium in the absence of mitochondria or substrate (succinate or glutamate). When mitochondrial suspensions (A) or substrate (B) were added, some electropolarization changes occurred

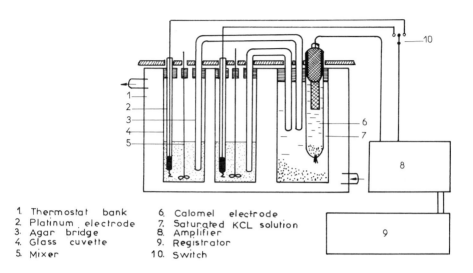

1. Thermostat bank
2. Platinum electrode
3. Agar bridge
4. Glass cuvette
5. Mixer
6. Calomel electrode
7. Saturated KCL solution
8. Amplifier
9. Registrator
10. Switch

1 A scheme of the apparatus used for monitoring changes in mitochondrial redox potentials.

between control and test samples which reached 50 to 60 mV, increasing at the metabolically active side. The extents of these changes were considerably diminished by metabolic poisons or by anaerobiosis. These results support the notion that changes in metabolic activity lead to potential differences and that their extent is proportional to the metabolic level. In general, polarization-induced intercellular potentials, measured by microelectrodes which are identical in principle to our electrodes, might represent a summation of cellular metabolic changes and membrane ionic processes.

Some of our own and some foreign studies have indicated that ACh shows nonmediator activities with respect to its concentration, i.e., stimulation or inhibition of oxygen consumption, of ATP-ase activity, and of ^{14}C-labeled amino acid incorporation into cell-free, protein-synthesizing systems. Stimulatory concentrations were regarded as optimal, whereas the inhibitory ones were regarded as minimal. Figure 3 shows that ACh produced a sharp decrease of electropolarization followed by a slow restoration of the initial level.

Figure 4 shows the effect of ACh in a concentration of $1 \times 10^{-3}M$

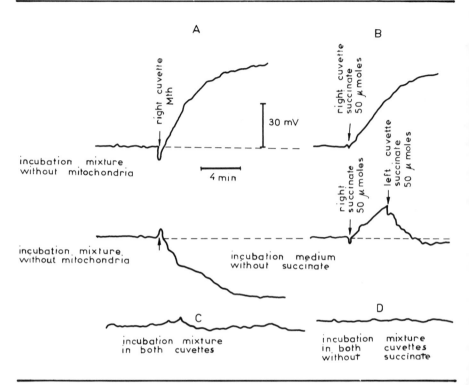

2   Changes in electrocharge of mitochondria depending on their metabolic states.

when the incubation medium was enriched by purified brain cholinesterase with an activity of 484 mg ACh hydrolized/min (B), as compared to control (A). When eserine was introduced at a concentration of 1×10^{-5} g/ml (C), the cholinesterase activity was 1.25 mg ACh hydrolized/min, while the control cholinesterase activity was 16.2 mg ACh hydrolized/min. It is evident that the augmented cholinesterase activity caused an increase in amplitude and a reduction in reversal time, whereas the addition of an anticholinesterase agent reduced both the amplitude and the restoration process.

The data regarding the effects of ACh on brain mitochondrial metabolic activity can be applied to explain further the molecular mechanism of regulation of brain function. In another series of experiments, using platinum and calomel electrodes, we examined the redox states of mitochondrial suspensions under normal conditions and when influenced

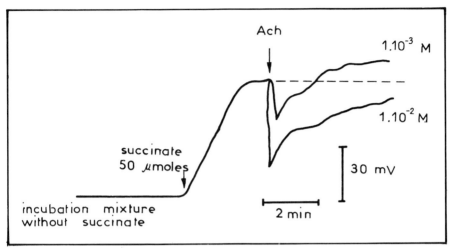

3 Characterization of changes in electropolarization potential in mitochondrial suspensions produced by addition of ACh.

Changes in the electrochemical potentials of mitochondrial suspensions caused by addition of ACh. A. Incubation medium, normal cholinesterase activity of 16.2 mg ACh hydrolyzed/min. B. Medium enriched in brain cholinesterase activity, 484 mg ACh hydrolyzed/min. C. Eserine was added at a concentration of 1×10^{-5} g/ml, cholinesterase activity of 1.25 mg ACh hydrolyzed/min.

4

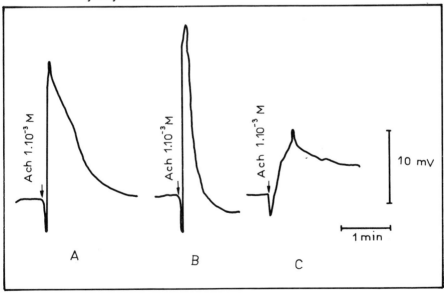

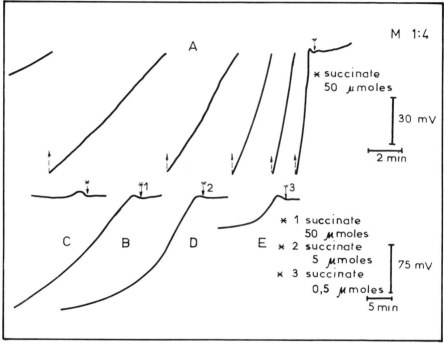

5   Redox changes in brain mitochondrial suspensions. A different intensity of the process was observed which depended upon the substrate concentration. The descending curves show increases of electrode potential which level off at substrate depletion.

by different concentrations of ACh and GABA. The experiments were carried out in a thermostatic glass cuvette at 30°C. Figure 5 shows voltage changes of the electrochemical potential in mitochondrial suspensions. The rate of this process can be calculated at short intervals in mV/sec, and the absolute values of changes can be expressed by means of the calibre signal given beforehand.

In another series of experiments, we demonstrated a relationship between the extent of changes and oxygen concentration. A system, (C), deprived of dehydrogenases and electron-seeking and electron transport systems, did not show any of the above-mentioned changes. An equivalent quantity of albumin or thermally denatured mitochondria was added in these tests. When ACh was added in concentrations of 1×10^{-5} M, 1×10^{-4} M, 1×10^{-3} M, and 1×10^{-2} M, quick peaks and

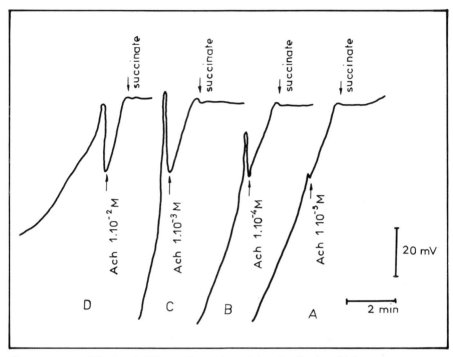

6 Effects of ACh on the redox state of mitochondrial suspensions. When ACh was present at 1 x 10^{-5} *M*, it was rather inactive; at concentrations of 1 x 10^{-4} *M* and 1 x 10^{-3} *M*, it caused a rise of peak amplitude and an increase in the rate of the following oxidizing process. The effect of ACh at a concentration of 1 x 10^{-2} *M*, had a greater time-peak continuity.

slow restoring changes (from 2 to 40 mV) were observed (Fig. 6).

Figure 7 shows the characteristics of changes which occurred when ACh was added together with the substrate. ACh, at 1×10^{-5} *M*, accelerated the process and at 1×10^{-2} *M*, it retarded the process (B). Also, when the rate of the electrochemical process was high, GABA decreased its intensity (D). This effect was clearer with a higher concentration of GABA (1×10^{-2} *M*). When there was a lower intensity process, GABA (at 1×10^{-3} *M* and 1×10^{-4} *M*) accelerated the redox changes.

Adrenaline caused some changes, which were reflected by a rise in redox potential.

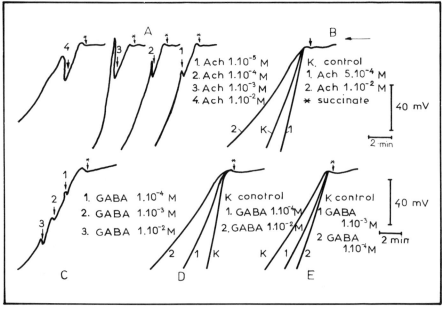

7   Effects of ACh and GABA on redox processes in mitochondrial suspensions.

As is well known, the redox potential of any system can be defined by the Nernst equation:

$$E = E^0 + \frac{RT}{nF} \ln \frac{(A_{ox})}{(A_{red})}.$$

Owing to the extreme heterogenity and complexity of the system investigated, three substances (ACh, GABA, and Adr), in different concentrations, were used in pure physicochemcial redox systems with different basic potentials. Differences were observed depending on the redox potential level and the ratios of oxidized to reduced forms of the components. The extents of the changes depended on the concentration. (A more extensive analysis of these processes is to be published.) Evidently the value and direction of the redox potential depend on the level of basic potential and on the power of the oxidizer or reducer. Some of the biologically active (redoxactive) substances can act by regulating changes of current redox state in microstructures at certain microintervals. The data obtained provide a possibility for a wider interpretation of the influence of mechanisms of biologically active

substances in biological heterogeneous function at the quantum molecular level.

D. SUMMARY

Mitochondria are among the most reactive of cellular structures
that respond to changes in incubation medium and other conditions by
altering their morphology, metabolism, and localization. We studied the
electrochemical activity and dynamics of redox potentials of brain
mitochondrial suspensions in relation to experimentally modulated
metabolism (i.e., alterations in the concentrations and types of substrates; e.g., ADP, oxygen, ATPase activity, metabolic poisons) and
the actions of different concentrations of acetylcholine, epinephrine, and
γ-aminobutyric acid on these phenomena. The speed and amplitude
characteristics of the changes obtained under diverse conditions were
determined. It is proposed that neuronspecific functions (biopolarization
and bioelectric activity) depend to a great extent on mitochondrial
activity and that neuromediators and some other biologically active
substances are metabolic modulators.

E. REFERENCES

Barreth, J., and Beis, I. (1973). The redox state of the free nicotine-amide-
    adenine dinucleotide couple in the cytoplasm and mitochondria of muscle
    tissue from *Ascaris lumbricoides* (Nematoda). *Comp. Biochem. Physiol.*
    44A, 331–340.
Beers, W. H., and Reich, E. (1970). Structure and activity of acetylcholine.
    Nature (London). 228,
Chynoweth, K. R., and Ternai, B. (1973). Nuclear magnetic resonance
    studies of the conformation and electron distributions in nicotine and
    in acetylcholine. *Molec. Pharmacol.* 9, 144–151.
Dutton, P. L., and Wilson, D. F. (1974). Redox potentiometry in mito-
    chondrial and photosynthetic bioenergetics. *Biochim. Biophys. Acta*
    346, 165–212.
Herman, Z. S., Kmiesiak-Kotada, K., Stominska-Znerek, J., and Szkilnik, R.
    (1972). Central effects of acetylcholine. *Psychpharmacologia* 27, 223–232.
Ivanov, R. I., Georgieva, N. D., and Kalchev, L. A. (1974). Changes in the
    redoxpotential of brain mitochondria and some physicochemical sys-
    tems induced by acetylcholine, γ-aminobutyric acid and adrenalin.
    Abstracts of II Congress of the Society for Physiological Sciences. Sofia,
    p. 67.
Kalchev, L. A. (1973). Regulation of mitochondrial processes forming bio-

electrical currents and substratum of memory. In: *Abstracts of 9th International Congress of Biochemists.*

Lovtrup, S., and Zelander, T. (1962). Isolation of brain mitochondria. *Exp. Cell. Res.* 27, 468.

Nachmansohn, D. (1970). Proteins in Bioelectricity. In: *Protein Metabolism of the Nervous System.* N.Y.–L., 313–333.

Pullman, A. (1965). *Molecular Biophysics.* Academic Press, New York.

Puppi, A. Szalay, L., and Dely, M. (1975). Interactions between the redox state of the biophase and the effect of acetylcholine on the activity of Na^+–K^+) ATP-ase in *Rana esculenta. Comp. Biochem. Physiol.* 50C, 75–79.

Tanaka, R. (1962). On the adenosine phosphatase activity of rat brain mitochondria. I. *Biol. Chem.* 237, 299.

9

Effects of Ethanol on Central Nervous System Cells[1]

ERNEST P. NOBLE

Neurochemistry Laboratory
Department of Psychiatry and Human Behavior
University of California
Irvine, California
and
Centre de Neurochimie du C.N.R.S.
and
Institute de Chimie Biologique
Faculte de Medecine
France

A. INTRODUCTION

Alcohol has been used and abused since time immemorial. Today in the United States, 95 million individuals drink alcohol and of this number 9 million are considered to be alcoholic persons.[2] Alcohol plays a major role in half of the highway fatalities, and of the 30,000 lives thus lost each year, more than half are youth between the ages of 16

[1]This work was conducted in collaboration with P. Syapin, V. Stefanovic, R. Massarelli and A. Rosenberg. I am grateful for the valuable encouragement and support of P. Mandel and G. Vincendon at the Centre de Neurochime, Strasbourg, France. The study was supported in part by a Guggenheim Fellowship and grants from the USPHS, National Institute on Alcohol Abuse and Alcoholism, ADAMHA (AA-00252); Thématique Programmée and Haut Comité de l'Alcoolism.

[2]First Special Report to the U.S. Congress on Alcohol and Health from the Secretary of Health, Education and Welfare, December, 1971.

and 24. Public intoxication alone accounts for one-third of all arrests reported annually, and alcohol was found in the body fluids of one-third of all homicide victims. Among certain segments of the United States population the incidence of alcoholism is at an epidemic level. For example, in some American Indian reservations, the rate of alcoholism is as high as 25 to 50%. The economic cost associated with misuse of alcohol is estimated at 25 billion dollars a year,[3] not to mention the psychological cost paid in the tragedy of shattered lives. Yet, the underlying causes of alcoholism remain enigmatic and treatment is often empirical and lacks a strong rational basis.

People drink alcohol (ethanol) for a variety of reasons, the most common being to induce a change in consciousness. Thus, the acute effects of ethanol on the CNS are well recognized in the inebriated subject. Chronic consumption of ethanol leads to behavioral changes as manifested by the severe symptoms of the withdrawal reaction and also by profound and long-lasting disorders in brain function. To gain an understanding into the basic mechanisms of these effects, increasing research effort has recently been focused on the neurochemical actions of ethanol using animal models. Despite some interesting and clearcut findings, the literature is disappointingly replete with contradictory or negative data (Noble and Tewari, 1976). The prevailing difficulty may be attributed to a number of technical or conceptual problems. Among some of the contributing factors may be the various species and strains of animals compared and the different doses and routes of ethanol employed in these studies. Furthermore, too frequently the brain has been treated as a homogeneous organ, which ignores its richness and complexity, thus obscuring any important effect that ethanol may exert on small but discrete systems.

Neural cells grown in culture offer several distinct advantages over the use of whole animals for studying the effects of alcohol on the CNS. Among these, the existence of clonal cell lines of glial and neuronal origin makes it possible to study separately the influence of ethanol on each cell type. Such studies are difficult to conduct and interpret with whole animal brain because fractionation and isolation procedures do not yield homogeneous populations of glia and neurons, and problems of cross-contamination as yet remain unresolved. Furthermore, culturing procedures make it feasible to control more carefully the dose of ethanol used and the environment for growth, and ethanol-induced effects on circulatory, hormonal, and other physiological variables, which typify

---

[3]Second Special Report to the U.S. Congress on Alcohol and Health from the Secretary of Health, Education and Welfare, June, 1974.

whole animal studies and which may confound the system under study, are eliminated.

The present report is apparently the first study dealing with the effects of ethanol on mammalian neural cells grown in culture. This work was begun at the Centre de Neurochimie, Strasbourg, France. It deals with three selected aspects of neural membranes as affected by acute and chronic exposure to ethanol. (1) Surface membrane topography as determined by the availability of sialic acid to hydrolytic cleavage, (2) the high affinity uptake of choline, and (3) the activity of membrane-associated enzymes, particularly those ectoenzymes found on the outer surface of the cell membrane.

B. MEMBRANE TOPOGRAPHY

Most of the sialic acid of the cell is found in its surface membrane (Wallach and Eylar, 1961; Lapetina et al., 1967; Glick et al., 1971), and since this membrane constitutes no more than 5% of the total mass of the cell, the actual concentration of sialic acid per milligram membrane protein is correspondingly high (Warren, 1976).

Glycoproteins and sialolipids are the predominant species containing sialic acid. This acid is generally found in terminal positions at the nonreducing ends of oligosaccharides which, in turn, are attached to proteins or lipids. Because of its prominent position in the membrane, the availability of sialic acid to hydrolytic cleavage can serve as a useful marker of alterations induced by various agents in membrane topography. The present study reports on externally added neuraminidase activity toward surface sialic acid of NN hamster astroblasts following acute and chronic exposure to ethanol. Methods for cell growth, incubation conditions and assay of sialic acid have been described elsewhere (Noble et al., 1975b, 1976.)

When intact NN cells were placed acutely in ethanol concentrations of up to 200 mM, neuraminidase-induced liberation of sialic acid was not altered indicating that at the concentration range used, ethanol did not influence the availability of sialic groups to cleavage by the exogenously added enzyme. However, cells grown in medium containing 100 mM ethanol yielded significant increases in neuraminidase-induced sialic acid release when compared to cells grown in the absence of the drug (Table I). It is of interest to note that the increased release occurred despite no significant differences in total cellular sialic acid between the control cells and cells grown in the presence of ethanol.

Analysis of cell components showed that the sialolipid fraction of both the control and chronic ethanol-treated cells was immune to

Table I. Neuraminidase Released Sialic Acid from Intact NN Cells Chronically Exposed to Ethanol[a]

Exposure to ethanol (days)	μg Sialic acid released/mg protein/30 min		
	Control	Ethanol exposed	p
19	2.41 ± 0.09 (5)	2.89 ± 0.06 (5)	< 0.005
66	2.27 ± 0.11 (6)	2.75 ± 0.07 (6)	< 0.01
88	3.07 ± 0.12 (6)	4.66 ± 0.16 (6)	< 0.001
110	2.99 ± 0.01 (5)	3.30 ± 0.09 (5)	< 0.05

[a] Cells were grown in the presence or absence of 100 mM ethanol in their media for the days indicated. After ethanol was removed the cells were incubated at 37°C with *Cl perfringens* neuraminidase and the amount of sialic acid released into the media was measured. Values are means ± S.E.M.; numbers of samples in parentheses; statistical significance is based on student's t-test.

attack by neuraminidase (Table II). This indicated that sialolipids of the intact cell surface appear to be more deeply localized so as to be sterically inaccessible to neuraminidase cleavage. This notion is supported by other studies (Weinstein et al., 1970; Barton and Rosenberg, 1973) using mammalian cell lines grown in culture.

When the glycoprotein fraction was analyzed following the addition of neuraminidase to intact cells, a large portion of its sialic acid was found to be reduced in the control as well as in the ethanol-grown cells (Table II). Furthermore, 10% more sialic acid in the glycoprotein fraction was released following neuraminidase treatment in cells chronically exposed to ethanol than in the control cells. This finding suggests that the observed enhancement of sialic acid release from the ethanol-treated cells by the hydrolytic enzyme is derived predominantly, if not exclusively, from their more accessible cell surface glycoproteins.

C.   CHOLINE UPTAKE

Choline is of fundamental importance in the CNS not only because of its role in the synthesis of acetylcholine but also because it is a component of phospholipids and as such plays a critical role in membrane function.

The uptake of choline has been investigated in several tissues, cells, or subcellular components (Martin, 1968; Flower et al., 1972; Yamamura and Snyder, 1973; Massarelli et al., 1974a). It has been shown that a carrier-mediated process, dependent on the ionic balance

Table II. Distribution of Sialic Acid in Various Fractions of NN Cells Chronically Exposed to Ethanol and Treated with Neuraminidase[a]

Cell fraction	Neuraminidase treatment	μg Sialic acid / mg protein	
		Control	Ethanol exposed
Glycoproteins	+	1.52 ± 0.37[b]	1.73 ± 0.15[c]
Glycoproteins	−	2.25 ± 0.16[b]	2.83 ± 0.12[c]
Sialolipids	+	1.34 ± 0.12[d]	1.43 ± 0.06[e]
Sialolipids	−	1.35 ± 0.15[d]	1.44 ± 0.09[e]
Total		3.83 ± 0.21	3.90 ± 0.10

[a] Cells were chronically exposed to 100 mM ethanol for 19 days. Intact cells were incubated at 37°C with **Cl. perfringens** neuraminidase for 30 min and cell fractions were obtained and analyzed for sialic acid. Values represent means ± S.E.M. for 5 samples; statistical significance is based on student's t-test.
[b] $P < 0.01$.
[c] $P < 0.001$.
[d, e] $P > 0.05$.

in the cellular environment, is involved in this phenomenon. Two major uptake systems of choline, differing in their affinity for substrate, have been described for neurons. It has been suggested that the high affinity system ($K_m = 10^{-6}M$) is directed toward the synthesis of acetylcholine (Yamamura and Snyder, 1973), whereas the general metabolism of choline depends most likely on the low affinity system. However, studies by Massarelli et al. (1974b) and Haeffner (1975) have shown that the high affinity uptake mechanism is not an exclusive property of neurons since glial cells or Ehrlich ascites cells grown in culture also demonstrate high and low affinity components.

In the present study the effects on the high affinity uptake of choline by neural cells exposed acutely or chronically to ethanol have been determined. A variety of cell lines have been used, including hamster astroblast clonal line NN, rat glioblast clonal line C_6, mouse neuroblastoma C 1300 adrenergic clone M_1, and cholinergic clone S_{21}. A preliminary report has already appeared (Noble et al., 1975a) and methods and detailed results appear elsewhere (Massarelli et al., 1976).

As in the case of surface sialic acid availability, the high affinity uptake of choline was not affected by the acute presence of 100 mM ethanol in the four cell lines studied. However, cells grown in the presence of 100 mM ethanol showed significant differences in choline uptake when compared to cells grown in the absence of this drug. More specifically, after 13 days of ethanol treatment, C_6 cells accumulated choline at a higher rate than control cells. A large increase in choline uptake was found at 15 days which then steadily decreased reaching control values after 41 days of treatment. Similarly, chronic ethanol exposure increased the high affinity uptake of choline in clone NN; however, this increase did not appear until 34 days of contact with ethanol and then returned toward control values as found in clone C_6.

To evaluate whether the effect of chronic ethanol exposure on the high affinity uptake of choline is limited to glial cells or if neuroblasts are also affected by chronic ethanol treatment, S_{21} cells were chronically exposed to 100 mM ethanol. After only 7 days in ethanol, the uptake of choline in clone S_{21} was increased by 40% over control cells. It appears, therefore, that neuroblast and glioblast cell lines are affected by ethanol. These findings indicate that chronic ethanol exposure affects not only a structural but also a functional characteristic of neural membranes.

D. MEMBRANE-ASSOCIATED ENZYMES

1. Adenosine Triphosphatase (ATPase) Activity

The acute and chronic effects of ethanol on ATPase activity in the brain has been the subject of numerous studies. Inhibition of Mg^{2+} ATPase and $(Na^+ + K^+)$ ATPase has been reported after the acute exposure to ethanol (Jarnefelt, 1961; Israel and Salazar, 1967; Sun and Samorajski, 1970). Chronic treatment with ethanol increased $(Na^+ + K^+)$ ATPase and Mg^{2+}ATPase activity differentially in regions of the cat brain (Knox et al., 1972). In rats, chronic ethanol treatment resulted in enhanced activity of $(Na^+ + K^+)$ATPase but not of Mg^{2+}ATPase activity (Israel et al., 1970; Roach et al., 1973). In the mouse, Israel and Kuriyama (1971) reported increased Mg^{2+}ATPase activity in the mitochondrial fraction after chronic ethanol treatment; however, no effect was observed on the microsomal or synaptosomal fractions. Goldstein and Israel (1972) found no effect on the $(Na^+ + K^+)$ATPase activity in brain homogenates of physically dependent mice, either immediately or 10 hr after ethanol withdrawal, while Mg^{2+}ATPase activity was lowered after a 10-hr withdrawal period.

The available data, therefore, indicate that ethanol presented in acute doses leads to inhibition of brain ATPase activity; however, chronic ethanol administration results in inconsistent effects on this enzyme system. Numerous factors may contribute to these disparate findings. For example, since isolated glia and neurons differ in their ATPase activity (Medzihradsky et al., 1971; Hamberger et al., 1975), it is not surprising that discordant results are obtained when data are compared among whole brain homogenates, or preparations of microsomes, mitochondria, or synaptosomes which contain varying amounts of material derived from neuronal and glial cells.

To ascertain more clearly and specifically if chronic exposure of CNS cells to ethanol alters their ATPase activity we have examined the effects of acute and chronic ethanol exposure on Mg^{2+}ATPase and $(Na^+ + K^+)$ATPase activities of NN hamster astroblasts and two mouse neuroblastoma clones, N_{IE-115} and M_1. A preliminary report has already been presented (Syapin et al., 1975) and more detailed methods and findings will appear elsewhere (Syapin et al., 1976a).

The activities of the two ATPase systems in the acute presence or absence of ethanol are shown in Table III. The NN cells showed inhibition of both Mg^{2+}ATPase and $(Na^+ + K^+)$ATPase activities at an ethanol concentration of 100 mM. Mg^{2+}ATPase activity decreased approximately 10%, while $(Na^+ + K^+)$ATPase activity was reduced to approximately 60% of control activity. The activities of these enzymes remained unchanged in the neuroblastoma cells at the same concentration of ethanol. The astroblast cells displayed higher $(Na^+ + K^+)$ATPase activity than the neuroblastoma cells. These findings are consistent with those of Kimelberg (1974) and Stefanovic et al (1975a).

The effects of prolonged exposure to 100 mM ethanol of hamster astroblasts and mouse neuroblasts on ATPase activity are shown in Table IV. Following 8 days in ethanol, the NN cell line showed a 28% increase in Mg^{2+}ATPase activity; however, $(Na^+ + K^+)$ATPase activity remained essentially unchanged. In neuroblastoma clone M_1 cells neither of the ATPase activities was affected by chronic treatment. In contrast, neuroblastoma clone N_{IE-115} cells showed stimulation in the activity of both enzymes.

Increased Mg^{2+}ATPase activity observed in the NN cells was found consistently over a wide time range of chronic ethanol exposure. Cultures were maintained in the presence of ethanol and tested for ATPase activity after 4, 6, 8, and 68 days of chronic exposure. The increased Mg^{2+}ATPase activity was the same whether cells were exposed to ethanol for 4 days or 68 days. $(Na^+ + K^+)$ATPase activity, however, did not change even after 68 days of exposure to ethanol.

When the phase of cell growth was taken into consideration, maximum increase in Mg^{2+}ATPase was found after 5 days of growth. However, the increased activity of this enzyme in the ethanol-treated cultures was independent of the phase of cell growth.

A number of explanations may be offered for the findings on chronic ethanol exposure. A plausible one is that prolonged exposure to the drug causes conformational changes in a component of the ATPase

Table III. Effect of Acute Exposure of Neural Cell Homogenates to Ethanol on their Mg^{2+}ATPase and (Na^++K^+)ATPase Activities[a]

Cell line	Mg^{2+}ATPase activity (μmole Pi liberated/mg protein/hr)		(Na^++K^+)ATPase activity (μmole Pi liberated/mg protein/hr)	
	Ethanol-free	100 mM Ethanol	Ethanol-free	100 mM Ethanol
NN	2.17 ± 0.08 (5)	1.90 ± 0.05 (5) $P < 0.02$	1.54 ± 0.13 (5)	0.96 ± 0.14 (5) $P < 0.02$
M_1	3.22 ± 0.29 (4)	2.89 ± 0.13 (4) NS	0.25 ± 0.04 (4)	0.21 ± 0.07 (4) NS
N_{IE}-115	5.55 ± 0.20 (5)	5.16 ± 0.18 (5) NS	0.28 ± 0.10 (5)	0.21 ± 0.08 (5) NS

[a] Cell cultures were obtained at 4 days growth. Values are mean ± S.E.M.; number of samples is given in parentheses; significance is based on student's t-test; NS = $P > 0.05$.

Table IV. Effect of Chronic Ethanol Exposure of Neural Cells on their Mg^{2+}ATPase and $(Na^{+}+K^{+})$ATPase Activities[a]

Cell line	Mg^{2+}ATPase activity (μmole Pi liberated/mg protein/hr)		$(Na^{+}+K^{+})$ATPase activity (μmole Pi liberated/mg protein/hr)	
	Ethanol-free	100 mM Ethanol	Ethanol-free	100 mM Ethanol
NN	3.08 ± 0.27 (5)	3.94 ± 0.02 (5) $P < 0.05$	1.61 ± 0.10 (5)	1.71 ± 0.14 (5) NS
M_1	3.33 ± 0.08 (5)	3.45 ± 0.17 (5) NS	0.15 ± 0.03 (5)	0.18 ± 0.04 (5) NS
N_{IE}-115	4.48 ± 0.18 (5)	5.98 ± 0.42 (5) $P < 0.05$	0.29 ± 0.08 (5)	0.62 ± 0.27 (5) $P < 0.05$

[a] Cultures were maintained either in a 100 mM ethanol-containing medium or normal medium (ethanol-free) continually for 8 days with one passage of cells after 4 days growth. Values are mean $\pm$ S.E.M.; numbers of samples in parentheses; significance is based on student's t-test; NS = $P > 0.05$.

system or in an adjacent membrane structure leading to a functional alteration in enzyme activity. Hokin et al. (1973) and Kyte (1972) have shown a major component of purified $(Na^+ + K^+)$ATPase preparations to be a glycoprotein rich in hydrophobic amino acids and sialic acid. It is plausible that ethanol, by virtue of its physicochemical characteristics, may alter the conformation of these components, thereby affecting ATPase activity. Indeed the studies on sialic acid reported earlier support the view that chronic ethanol exposure of cells results in steric modification in surface glycoproteins. Other aspects of neural membrane structure may also be influenced by ethanol. Dinovo et al. (1976) have found enhanced exposure to sulfhydryl groups in brains of rats chronically treated with ethanol. While alterations in membrane topography may be an important determinant of the enhanced activity of ATPase, the present data do not rule out an enzyme induction phenomenon by the drug.

2. 5'-Nucleotidase

5'-Nucleotidase is frequently found to behave as a plasma membrane enzyme in subcellular fractions of mammalian tissues (DePierre and Karnovsky, 1973). Evidence is accumulating to suggest that the active site of this enzyme faces the external medium rather than the cytoplasm (Trams and Lauter, 1974; DePierre and Karnovsky, 1974), and because of this characteristic the enzyme is referred to as an ectoenzyme. In view of the suggestive evidence provided from studies of sialic acid and choline uptake that ethanol induces changes in the plasma membrane, it was of interest to determine how this drug would affect the properties of an ectoenzyme. The effects of acute and chronic ethanol exposure on 5'-nucleotidase activity were assessed using four neural cell lines grown in culture: rat glioma clonal line C_6, hamster astroblast clone line NN, mouse C 1300 neuroblastoma clonal line M_1, and neuroblastoma clonal line $N_{IE^{-115}}$. The results have been presented in a preliminary report (Syapin et al., 1975). More detailed data will appear elsewhere (Syapin et al., 1976b).

Table V shows the effects of exposure of the four cell lines to 100 mM ethanol for 6 days. Significant increases in 5'-nucleotidase activity were found in the C_6, NN, and $N_{IE^{-115}}$ cells; however the M_1 cells remained essentially unaffected by ethanol. It is also of interest to note that the C_6 cells possessed the highest 5'-nucleotidase activity of all the cells tested. Indeed the activity in the C_6 cells was 2–12 times greater than those reported for rat fat cells (Newby et al., 1975), HeLa cells (Trams and Lauter, 1974; Johnsen et al., 1974), and human carcinoma

Table V. Effect of 6 Days Chronic Exposure to Ethanol on
Ecto-5'-Nucleotidase Activity in Various Clonal Cell Lines[a]

| Clonal cell lines | Ecto-5'-nucleotidase activity[b] | | |
	Ethanol-free	100 mM Ethanol	P
C_6	1.70 ± 0.29 (4)	2.63 ± 0.19 (4)	< 0.005
NN	0.200 ± 0.011 (5)	0.293 ± 0.007 (5)	< 0.001
M_1	0.103 ± 0.008 (5)	0.105 ± 0.008 (5)	NS
$N_{IE\text{-}115}$	0.079 ± 0.005 (5)	0.101 ± 0.003 (5)	< 0.001

[a] Cultures were grown in the presence or absence of 100 mM ethanol in their media for
6 days. Values are means ± S.E.M.; number of samples in parentheses; statistical
significance is based on student's t-test; NS = P > 0.05.
[b] Micromole Pi liberated/mg protein/hr.

KB cells (Trams and Lauter, 1974). Because of this consideration the
remainder of the studies deal only with the C_6 cells.

The effect of 6 days exposure of C_6 cells to increasing concentrations of ethanol was determined on 5'-nucleotidase activity. An interesting dose-response relationship was observed. The highest enzyme activity
was found when cells were grown in 10 mM or 150–400 mM ethanol,
whereas cells exposed to 20 mM or 50 mM ethanol had enzyme activities
that were not significantly different from control cells.

C_6 cells were also exposed to the acute action of increasing concentrations of ethanol and their effect on 5'-nucleotidase was determined.
5'-Nucleotidase activity increased at low (10–20 mM) concentrations of
ethanol, decreased at 50 mM ethanol, and increased again at high
(100–400mM) concentrations of the drug.

Since the dose-response characteristics of the acute and chronic
studies were similar, the possibility arose that the stimulation of 5'-
nucleotidase by prolonged ethanol exposure might be due to the
lingering presence of ethanol because of insufficient washing of the
cell monolayer with ethanol-free medium prior to enzyme assay. However, several experiments refuted this possibility. What remains unique
in these studies is that the acute presence of ethanol enhanced 5'-nu-
cleotidase activity whereas in the case of ATPase, inhibition on activity
was found as noted earlier. It is tempting to suggest that the ecto-
characteristic of 5'-nucleotidase may be the distinguishing factor in the
differential response to ethanol by these two enzymes; however, additional studies on other ectoenzymes are necessary before a firm conclusion is reached on this notion.

To elucidate the mechanisms involved in the stimulation of 5'-nucleotidase following chronic exposure to ethanol, further studies were conducted on C_6 cells grown in the presence of 100 mM ethanol. The activity of 5'-nucleotidase was determined over a pH range of 6.5 to 9.5. Control cells exhibited double pH optimum, one at pH 7.4 and the other at pH 9.0. Chronic ethanol-treated cells, while showing increased 5'-nucleotidase activity over the entire pH range tested revealed, nonetheless, similar double optimum as the control cells. The activity of 5'-nucleotidase toward different phosphate ester substrates was examined in intact cells, so as to obtain ectoenzymatic activity. Activity was also determined in cell homogenates, so as to determine total 5'-nucleotidase activity. Table VI shows that 5'-AMP was the most active substrate for control and chronic ethanol-treated cells. The relative rates of substrate hydrolysis for both these cells were 5'-AMP > 5'-CMP > 5'-GMP = 5'-IMP. This was true for ecto-5'-nucleotidase activity as well as total 5'-nucleotidase activity.

It is of particular interest to note that the differences between the ethanol-treated and control ecto-5'-nucleotidase activity (B–A) when compared to the differences between the ethanol-treated and control total 5'-nucleotidase activity (D–C) were remarkably similar for the four of the above specific substrates used. This finding would indicate that although nonecto-5'-nucleotidase activity is found in the whole C_6 cells, the ethanol enhanced 5'-nucleotidase activity occurred predominantly if not entirely in the ectoenzyme fraction.

Lineweaver-Burk plots were constructed for the 5'-nucleotidase of intact control cells and cells grown 6 days in 100 mM ethanol. K_m was decreased by 40% in the chronic ethanol-treated cells, whereas V_{max} decreased by only 7%. These observations would suggest that ethanol either induced an enhanced exposure of the enzyme catalytic site or increased the affinity of the enzyme for substrate.

Concanavalin A (Con A) is a protein isolated from Jack Bean which binds to specific sugars found at the cell surface (So and Goldstein, 1968). This binding powerfully inhibits ecto-5'-nucleotidase activity of intact C_6 cells, possibly through a reduction in the amount of active enzyme on the cell surface (Stefanovic et al., 1975b). When Con A was added to intact C_6 cells chronically exposed to 100 mM ethanol, ecto-5'-nucleotidase activity was reduced to the level of the control cells. That this effect on Con A was due to binding of the lectin to the cell surface was shown when the inhibitory effect of Con A was abolished with its removal by α-methyl-d-mannoside. It was of further interest that a double reciprocal plot analysis showed that ethanol was a non-

Table VI. Effect of Chronic Ethanol Treatment on Substrate Specifically of Ecto- and Total 5'-Nucleotidase Activity in C_6 Cells[a]

Substrate 3 mM	Ecto 5'-nucleotidase activity[b]			Total 5'-nucleotidase activity[b]		
	Control cells (A)	Treated cells (B)	(B-A)	Control cells (C)	Treated cells (D)	(D-C)
5'-AMP	1.70 ± 0.13	2.63 ± 0.10	0.93	3.13 ± 0.16	4.17 ± 0.17	1.04
5'-CMP	1.29 ± 0.09	1.64 ± 0.08	0.35	2.34 ± 0.10	2.59 ± 0.28	0.25
5'-GMP	0.76 ± 0.01	1.03 ± 0.03	0.27	1.23 ± 0.23	1.46 ± 0.83	0.23
5'-IMP	0.77 ± 0.01	1.11 ± 0.03	0.34	1.33 ± 0.13	1.70 ± 0.20	0.37
5'-dAMP	0.31 ± 0.05	0.33 ± 0.12	0.02	1.02 ± 0.10	2.24 ± 0.15	1.22
p-Nitro-phenyl-phosphate	0.19 ± 0.01	0.21 ± 0.02	0.02	1.18 ± 0.05	0.17 ± 0.07	-0.47
G-6-P	0.05 ± 0.01	0.09 ± 0.01	0.04	—	—	—

[a] Cells were grown in the presence of 100mM ethanol in their media for 6 days. Values are means ± S.E.M. of 3 determinations for each substrate. Ectoenzymatic activity was measured on intact attached cells and total activity was measured in cell homogenates.
[b] Micromole Pi liberated/mg protein/hr.

177

competitive antagonist of Con A suggesting that ethanol is acting at a site in the membrane distant from that of Con A.

E. SUMMARY AND CONCLUSIONS

The acute and chronic actions of ethanol have been determined on neural cells grown in culture. Five clonal lines have been used which include: NN hamster astroblast, C_6 rat glioma, M_1 and N_{IE^-115} mouse adrenergic neuroblastoma, and S_{21} mouse cholinergic neuroblastoma cells. Three selected characteristics of neural membranes have been investigated: surface-membrane topography, the high affinity uptake of choline, and activity of membrane-associated enzymes.

Because of its prominent position in the membrane, sialic acid availability to hydrolytic cleavage by neuraminidase has been used as a marker of alterations induced by ethanol in membrane topography. Ethanol concentrations of up to 200 mM added acutely to intact NN cells failed to influence the enzymatic release of sialic acid. However, when these cells were exposed for prolonged periods to 100 mM ethanol, significant increases were found in neuraminidase-induced sialic acid release. This occurred despite no significant changes in total cellular sialic acid. Analysis of cell components showed no changes in the sialolipid fraction, whereas in the glycoproteins, neuraminidase treatment markedly reduced sialic acid content with the percent reduction being greater in the cells chronically exposed to ethanol than controls. These data suggest that exposure of cells for prolonged periods to ethanol results in steric modification of surface glycoproteins.

The high affinity uptake of choline was not altered in NN, C_6, M_1, and S_{21} cells by acute exposure to 100 mM ethanol. However, prolonged exposure to 100 mM ethanol resulted in an increase in the high affinity uptake of choline, with the earliest effects shown by the S_{21} neuronal cells followed in decreasing order by the two glial cells, the C_6 and the NN.

The effects of acute and chronic ethanol exposure on the activity of Mg^{2+}ATPase and $(Na^+ + K^+)$ATPase, two membrane-associated enzymes, have been determined. Acute exposure of NN cell homogenates to 100 mM ethanol inhibited the activity of both enzymes, whereas no significant inhibition of these enzymes was observed in the neuroblastoma cells M_1 and N_{IE^-115}. Prolonged contact of NN cells with 100 mM ethanol increased the activity of only Mg^{2+}ATPase, with no significant effects being observed on both enzymes in the M_1 cells. When N_{IE^-115} cells were exposed chronically to ethanol, enhanced activities were obtained for Mg^{2+}ATPase and $(Na^+ + K^+)$ATPase.

5′-Nucleotidase, an enzyme found on the surface membrane (ecto-enzyme) as well as in other structures of the cell, was studied. Acute exposure to ethanol, in contrast to the effects on ATPase, stimulated ecto-5′-nucleotidase activity. Exposure to 100 mM ethanol for 6 days resulted in increased ectoenzymatic activity of C_6, NN, N_{IE-115}, but not M_1 cells. Detailed biochemical characteristics of 5′-nucleotidase were studied in C_6 cells following chronic exposure to 100 mM ethanol. No shifts occurred in the double pH optimum peaks of this enzyme. 5′-AMP was the most active substrate for control and chronic ethanol-treated cells. The enhanced activity following chronic ethanol exposure occurred predominantly if not entirely in the ecto-5′-nucleotidase fraction. Kinetic analysis showed a 40% decrease in K_m and only a 7% decrease in V_{max} of the cells chronically treated with ethanol. Concanavalin A abolished the enhanced stimulation of ecto-5′-nucleotidase activity that occurred after prolonged ethanol exposure.

In summary, neural cells have been shown to be useful models for elucidating the mechanism of action of ethanol on the central nervous system. Using structural, functional, and enzymatic criteria, evidence has been provided which indicates that ethanol exerts critical effects on membranes particularly those found on the cell surface.

F. REFERENCES

Barton, N. W., and Rosenberg, A. (1973). Action of *Vibrio cholerae* neuraminidase (sialidase) upon the surface of intact cells and their isolated sialolipid components. *J. Biol. Chem.* 248: 7353–7358.

DePierre, J. W., and Karnovsky, M. L. (1973). Plasma membranes of mammalian cells. A review of methods for their characterization and isolation. *J. Cell Biol.* 56: 275–303.

DePierre, J. W., and Karnovsky, M. L. (1974). Ecto-enzymes of the guinea pig polymorphonuclear leukocyte. I. Evidence for an ecto-adenosine monophosphatase-adenosine triphosphatase, and -*p*-nitrophenol phosphatase. *J. Biol. Chem.* 249: 7111–7120.

Dinovo, E. C., Gruber, B., and Noble, E. P. (1976). Alterations of fast-reacting sulfhydryl groups of rat brain microsomes by ethanol. *Biochem. Biophys. Res. Commum.* 68: 975–981.

Flower, R. J., Pollitt, R. J., Sanford, P. A., and Smyth, D. H. (1972). Metabolism and transfer of choline in hamster small intestine. *J. Physiol. (London)* 226: 473–489.

Glick, M. C., Comstock, C. A., Cohen, M. A., and Warren, L. (1971). Membranes of animal cells. VIII. Distribution of sialic acid, hexosamines and sialidase in the L cell. *Biochem. Biophys. Acta* 233: 247–257.

Goldstein, D. B., and Israel, Y. (1972). Effect of ethanol on mouse brain

(Na+K)-activated adenosine triphosphatase. *Life Sci.* 11, Part II, 957–963.

Haeffner, E. W. (1975). Studies on choline permeation through the plasma membrane and its incorporation into phosphatidyl choline of Ehrlich-Lettré-ascites tumor cells *in vitro*. *Eur. J. Biochem.* 51: 219–228.

Hamberger, A., Babitch, J. A., Blomstrand, C., Hansson, H., and Sellström, A. (1975). Evidence for differential function of neuronal and glial cells in protein metabolism and amino acid transport. *J. Neurosci. Res.* 1: 37–56.

Hokin, L. E., Dahl, J. L., Deupree, J. D., Dixon, J. F., Hackney, J. F., and Perdue, J. E. (1973). Studies on the characterization of the sodium-potassium transport adenosine triphosphatase. *J. Biol. Chem.* 248: 2593–2605.

Israel, M. A., and Kuriyama, K. (1971). Effect of *in vivo* ethanol administration on adenosinetriphosphatase activity of subcellular fractions of mouse brain and liver. *Life Sci.* 10, Part II, 591–599.

Israel, Y., Kalant, H., LeBlanc, E., Bernstein, J. C., and Salazar, I. (1970). Changes in cation transport and (Na+K)-activated adensoine triphosphatase produced by chronic administration of ethanol. *J. Pharm. Exp. Ther.* 174: 330–336.

Israel, Y., and Salazar, I. (1967). Inhibition of brain microsomal adenosine triphosphatases by general depressants. *Archs Biochem. Biophys.* 122: 310–317.

Jarnefelt, J. (1961). Inhibition of the brain microsomal adenosinetriphosphatase by depolarizing agents. *Biochim. Biophys. Acta* 48: 111–116.

Johnsen, S., Stokke, T., and Prydz, H. (1974). HeLa cell plasma membranes. I. 5'-Nucleotidase and ouabain-sensitive ATPase as markers for plasma membranes. *J. Cell Biol.* 63: 357–363.

Kimelberg, H. K. (1974). Active potassium transport and (Na$^+$+K$^+$)ATPase activity in cultured glioma and neuroblastoma cells. *J. Neurochem.* 22: 971–976.

Knox, W. H., Perrin, R. G., and Sen, A. K. (1972). Effect of chronic administration of ethanol on (Na+K)-activated ATPase activity in six areas of the cat brain. *J. Neurochem.* 19: 2881–2884.

Kyte, J. (1972). Properties of the two polypeptides of sodium- and potassium-dependent adenosine triphosphatase. *J. Biol. Chem.* 247: 7642–7649.

Lapetina, E. G., Soto, E. F., and DeRobertis, E. (1967). Gangliosides and acetylcholinesterase in isolated membranes of the rat-brain cortex. *Biochim. Biophys. Acta* 135: 33–43.

Martin, K. (1968). Concentrative accumulation of choline by human erythrocytes. *J. Gen. Physiol.* 51: 497–516.

Massarelli, R., Ciesielski-Treska, J., Ebel, A., and Mandel, P. (1974a). Kinetics of choline uptake in neuroblastoma clones. *Biochem. Pharmacol.* 23: 2857–2865.

Massarelli, R., Ciesielski-Treska, J., Ebel, A., and Mandel, P. (1974b). Choline uptake in glial cell cultures. *Brain Res.* 81: 361–363.

Massarelli, R., Syapin, P. J., and Noble, E. P. (1976). *Life Sci.* 18: 397–404.

Medzihradsky, F., Nandhasri, P. S., Idoyaga-Vargas, V., and Sellinger, O. Z. (1971). A comparison of the ATPase activity of the glial cell fraction and the neuronal perikaryal fraction isolated in bulk from rat cerebral cortex. *J. Neurochem,* 18: 1599–1603.

Newby, A. C., Luzio, J. P., and Hales, C. N. (1975). The properties and extracellular location of 5'-nucleotidase of the rat-cell plasma membrane. *Biochem. J.* (London) 146: 625–633.

Noble, E. P., Massarelli, R., and Syapin, P. J. (1975a). The acute and chronic effects of ethanol on glial and neuronal cells in tissue culture. Sixth International Congress of Pharmacology, Helsinki, July, 1975, Abs. #243.

Noble, E. P., Syapin, P. J., Vigran, R., Gombos, G., Vincendon, G., and Rosenberg, A. (1975b). The effects of ethanol on neuraminidase releasable sialic acid of cultured neural cells. Fifth International Meeting of the International Society for Neurochemistry, Barcelona, Spain, September, 1975, Abs. #218.

Noble, E. P., Syapin, P. J., Vigran, R., and Rosenberg, A., (1976). Neuramindase-releasable surface sialic acid of cultured astroblasts exposed to ethanol. *J. Neurochem.,* in press.

Noble, E. P., and Tewari, S. (1976). Metabolic aspects of alcoholism in the brain. In: *Metabolic Aspects of Alcoholism* (C. Lieber, ed.). Medical and Technical Publishing Co., Lancaster, England, in press.

Roach, M. K., Khan, M. M., Coffman, R., Pennington, W., and Davis, D. L. (1973). Brain ($Na^+{+}K^+$)-activated adenosine triphosphatase activity and neurotransmitter uptake in alcohol-dependent rats. *Brain Res.* 63: 323–329.

So, L. L., and Goldstein, I. J. (1968). Protein–carbohydrate interaction. XX. On the number of combining sites on concanavalin A., the phytohemagglutinin of the Jack Bean. *Biochim. Biophys. Acta* 165: 398–404.

Stefanovic, V., Ciesielski-Treska, J., Ledig, M., and Mandel, P. (1975a). ($Na^+{+}K^+$)activated ATPase and K^+-activated p-nitrophenylphosphatase activities of the nervous system cells in tissue culture. *Experientia,* 31: 807–808.

Stefanovic, V., Mandel, P., and Rosenberg, A. (1975b). Concanavalin A inhibition of ecto-5'-nucleotidase of intact cultured C_6 glioma cells. *J. Biol. Chem.* 250. 7081–7083.

Sun, A. Y., and Samorajski, T. (1970). Effects of ethanol on the activity of adenosine triphosphatase and acetylcholinesterase in synaptosomes isolated from guinea-pig brain. *J. Neurochem.* 17: 1365–1372.

Syapin, P. J., Stefanovic, V., Ciesielski-Treska, J., Mandel, P., and Noble, E.P. (1975). The effects of ethanol on growth and on ecto-enzymatic activity of cultured cells. Fifth International Meeting of the International Society for Neurochemistry, Barcelona, Spain, September, 1975, Abs. #232.

Syapin, P. J., Stefanovic, V., Mandel, P., and Noble, E. P. (1976a). The chronic and acute effects of ethanol on adenosine triphosphatase activity in cultured astroblast and neuroblastoma cells. *J. Neuroscience Res.*, in press.

Syapin, P. J., Stefanovic, V., Mandel, P., and Noble, E. P. (1976b). In preparation.

Trams, E. G., and Lauter, C. J. (1974). On the sidedness of plasma membrane enzymes. *Biochim. Biophys. Acta* 345: 180–197.

Wallach, D. F., and Eylar, E. H. (1961). Salic acid in the cellular membranes of Ehrlich-ascites-carcinoma cells. *Biochim. Biophys. Acta* 52: 594–596.

Warren, L. (1976). In: *The Biological Roles of Sialic Acid* (A. Rosenberg, and C. L. Schengraund, eds.), Plenum Press, New York, in press.

Weinstein, D. B., Marsh, J. B., Glick, M. C., and Warren, L. (1970). Membranes of animal cells. VI. The glycolipids of the L cell and its surface membrane. *J. Biol. Chem.* 245: 3928–3937.

Yamamura, H. I., and Snyder, S. H. (1973). High affinity transport of choline into synaptosomes of rat brain. *J. Neurochem.* 21: 1355–1374.

Behavioral Neurochemistry

10

Cerebellar Cyclic Nucleotides, Prostaglandin E₂ and Some Convulsant and Tremorogenic Drugs[1]

R. FUMAGALLI
F. BERTI
G. C. FOLCO
D. LONGIAVE
R. PAOLETTI

Institute of Pharmacology and Pharmacognosy
University of Milan
Milan, Italy

A. INTRODUCTION

Prostaglandins (PGs), a family of endogenous acidic lipids present in mammalian brain and in peripheral tissues, are probably involved in the functions of the central nervous system (CNS), although their role is far from being elucidated. PGs, biosynthesized in rat brain cortex (Nicosia and Galli, 1975; Wolfe et al., 1976), have been found in brain homogenates rich in nerve endings (Kataoka et al., 1967). PGs are also released from the cerebral cortex (Ramwell and Show, 1966; Bradley et al., 1969), cerebellum (Coceani and Wolfe, 1965) and spinal cord (Coceani et. al., 1971). The regulatory role of PGs on intracellular concentrations of cyclic nucleotides in peripheral events (such as lipolysis, gastric secretion, luteolysis) (Ramwell and Shaw, 1970; Shaw et al., 1972; Kuehl, 1974) as well as in the CNS, is well estab-

[1]This experimental work has been partially supported by the Consiglio Nazionale delle Ricerche (C.N.R.), Rome, Italy, Contributo No. 74.00173.04

lished. Avanzino et al. (1966) have provided evidence that prostaglandins exert a direct action on central neurons, and later Hoffer et al. (1969) showed that the microelectrophoretic applications of PGE_1 and PGE_2 antagonize the reduction in discharge rate of rat Purkinje cells evoked by norepinephrine, but not by cyclic AMP (3′,5′-adenosine cyclic monophosphate). Additional support for the role played by PGs at this level has been provided by the histochemical localization of prostaglandin dehydrogenase in the Purkinje cell layer of cerebellum (Siggins et al., 1971). These workers have hypothesized that the antagonism between PGEs and norepinephrine occurs at the level of cyclic AMP formation. Further evidence for the ability of prostaglandins of the E type to interact with the cyclic AMP (cAMP) system of the CNS has been provided by *in vitro* studies carried out in our laboratory (Berti et al., 1973). It has been shown that the conversion of prelabeled ATP into cAMP in rat brain cortical slices is stimulated by PGE_1 and E_2, but not by $PGF_{2\alpha}$.

From the pharmacological point of view, prostaglandins of the E type elicit sedative and tranquilizing effects when administered to intact animals (Holmes and Horton, 1968). Moreover, PGE_1 exhibits anticonvulsant properties that depend on the experimental model used; e.g., this prostaglandin protects almost completely against pentamethylenetetrazole (PMT), but not against picrotoxin-induced convulsions (Horton, 1972). Extensor seizures due to maximal electroshock are only partially prevented by PGE_1, whereas conflicting results have been obtained in strychnine-induced convulsions (Duru and Türker, 1969; Horton, 1972). Recently, it has been reported that experimental convulsions greatly influence cerebellar concentrations of cAMP and/or cyclic-3′,5′-guanosine monophosphate (cGMP) and that the mechanisms of action of some convulsant and anticonvulsant drugs are at least partially correlated with variations of the cyclic nucleotides in the cerebellum (Goldberg et al., 1972; Mao et al., 1974, 1975; Costa et al., 1975). The well-established anticonvulsant properties of PGEs prompted us to investigate their possible interference with cyclic nucleotide concentrations that occur in the cerebellum during experimental convulsions. Similar experiments have been designed with harmaline, a drug that specifically induces cerebellar tremors and increases cerebellar concentrations of cGMP (Llinas and Volkind, 1973; Mao and Guidotti, 1974).

B. MATERIALS AND METHODS

Male Sprague-Dawley rats (C. Erba, Milan, Italy) weighing 140–160 g were used. All animals were housed for 1 week before the experiments and had free access to food and water.

1. Animal Sacrifice by Microwave Oven

Rats were sacrificed by exposure of their heads for 4 sec to a high intensity focused microwave radiation (70 mW/cm^2), generated by a magnetron (2.4 kW, 2.49 GH$_2$), in order to minimize postmortem variations of cyclic nucleotide concentrations (Uzunov and Weiss, 1971; Guidotti et al., 1974).

2. Assays of cGMP and cAMP

Cyclic GMP and cAMP were assayed in cerebellum after tissue homogenization in 0.4 N HClO$_4$, purification by column chromatography, and separation of the nucleotides following the procedure previously described by Mao and Guidotti (1974). Cyclic AMP was measured according to the method of Gilman (1970), using a binding protein from bovine skeletal muscle that was isolated and purified in our laboratory. The final incubation volume was 50 μl, and the sensitivity of the method allowed detection of 0.1 pmole of the nucleotide. Cyclic GMP was determined using the radioimmunoassay technique of Steiner et al. (1972). A commercially available kit for cGMP radioimmunoassay was purchased from Collaborative Res. Inc., Boston, Massachusetts, United States.

The cerebellar content of cyclic nucleotides is expressed as picomoles/milligram protein, the protein being measured by the method of Lowry et al. (1951), using bovine serum albumin (faction V) as standard.

3. Drugs

Pentamethylenetetrazole (PMT) was kindly supplied by Knoll Pharmaceuticals, Milan, Italy. Harmaline-HCl was purchased from Sigma, London, England. PGE$_2$ was generously supplied by Dr. K. Sano, Ono Pharmaceuticals, Osaka, Japan and by Dr. J. E. Pike, Upjohn Co., Kalamazoo, Michigan, United States.

C. RESULTS

1. PGE$_2$ and Pentamethylenetetrazole (PMT) Convulsions

Convulsions induced by PMT have been correlated with variations of cerebellar cyclic nucleotides. The choice of this drug is based mainly on the original finding by Horton (1972) of a protective action of PGEs against PMT-induced seizures. In our experiments the anticonvulsant activity of PGE$_2$ (2.8 μmoles/kg, i.p.) was maximal between 5 and 7 min after its administration.

Figure 1 shows the effects of a convulsant dose of PMT (100 mg/kg, i.p.) on rat cerebellar cyclic nucleotides. The concentrations of both

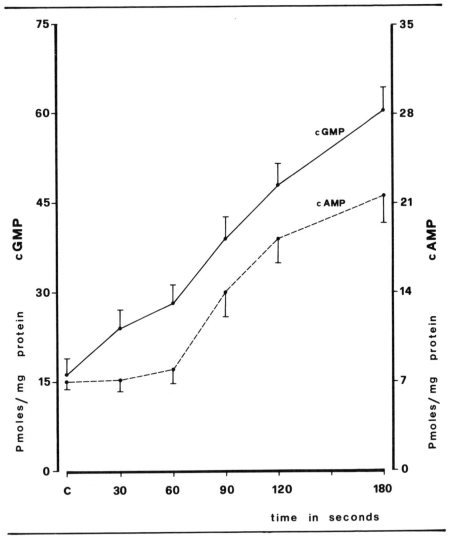

1 Time course of the effect of PMT on cerebellar cyclic nucleotides. PMT, 100 mg/kg, i.p. at zero time; means ± S.E.M. (8 rats in all cases).

nucleotides were significantly increased during convulsions, though their time profiles were rather different. Although the increase of cGMP was already evident at 30 sec, cAMP levels remained unchanged even 60 sec after PMT, and increased only after the onset of convulsions (60–90 sec). The same relationship holds between cerebellar cyclic nucleotides

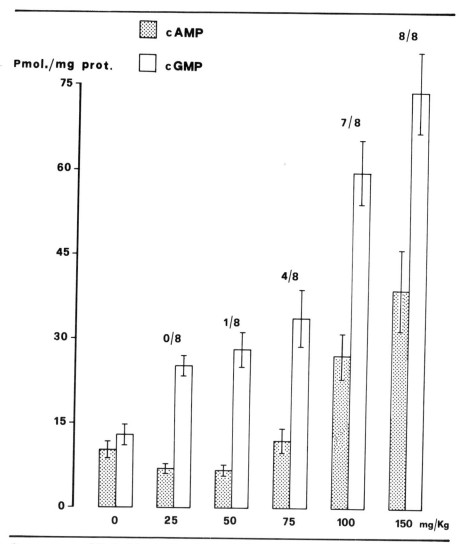

2 Effect of increasing doses of PMT on cerebellar cyclic nucleotides. PMT was given i.p. 90 sec before sacrifice. The numbers of animals convulsed by the treatment are provided above the bars.

and convulsions elicited by increasing doses of PMT (Fig. 2). Subconvulsant doses of PMT (25 and 50 mg/kg, i.p.) doubled the concentration of cGMP without affecting the levels of cAMP. Larger doses of PMT were followed by a higher incidence of convulsions, and a concomitant increase of both nucleotides.

Table I. Time Course of the Effect of a Subconvulsant Dose
of PMT on Cerebellar Cyclic Nucleotides

Treatment	cGMP[b] (pmoles/mg protein)	cAMP[b] (pmoles/mg protein)
Controls	14.10 ± 2.8	14.2 ± 2.0
PMT[a]		
30 sec.	22.06 ± 2.5[c]	10.3 ± 1.3
90 sec.	28.13 ± 3.1[d]	14.0 ± 2.7
180 sec.	27.06 ± 2.8	10.7 ± 1.8
360 sec.	22.20 ± 2.6	11.0 ± 2.7

[a] PMT:25 mg/kg i.p.
[b] Means $\pm$ S.E.M.; values from 8 rats in all cases.
[c] $P < 0.05$.
[d] $P < 0.01$.

The relationship between a subconvulsant dose of PMT (25 mg/kg, i.p.) and cerebellar cyclic nucleotides has been confirmed by further experiments. Table I provides a time course profile showing a prompt elevation of cGMP content 90 sec after PMT. In contrast cAMP levels were not affected.

The reported protective action of PGEs against PMT-induced seizures (Horton, 1972) has been evaluated both pharmacologically and biochemically and has been compared with that of chlordiazepoxide (CDP), an anticonvulsant benzodiazepine. PGE_2 (2.8 μmoles/kg, ip.), given 3.5 min before PMT (100 mg/kg, i.p.), completely prevented convulsions in 6 out of 8 rats and greatly reduced the duration and severity of convulsions in the remaining two animals. The increase of both cyclic nucleotides elicited by PMT (Table II) was fully prevented by PGE_2, but the same dose of $PGF_{2\alpha}$ was inactive. Chlordiazepoxide, on the other hand, completely inhibited the elevation of cAMP but only partially inhibited that of cGMP. Moreover, CDP, but not PGE_2, reduced the basal levels of cGMP. The influence of chlordiazepoxide on the cerebellar contents of both cyclic nucleotides is shown in Fig. 3. Basal values of cGMP, but not of cAMP, were significantly reduced; however, the levels of cGMP increased with time to values higher than those of untreated controls, but still were much lower than after PMT in the nonprotected group. In agreement with the above data (Table II) the cAMP level did not increase in the cerebella of the group that was protected from convulsions.

Table II. Cerebellar Cyclic Nucleotides: Effect of CDP and
of PGE$_2$ on PMT-induced Convulsions

Treatment[a]		cGMP[b] (pmoles/mg protein)	cAMP[b] (pmoles/mg protein)
Controls	(16)	8.6 ± 1.7	10.3 ± 1.2
PMT	(16)	57.3 ± 3.4[c]	27.9 ± 4.2[e]
CDP	(8)	2.8 ± 0.4[c]	8.1 ± 0.8
PGE$_2$	(8)	7.3 ± 1.1	10.1 ± 1.5
PMT + CDP	(8)	16.7 ± 1.8[c]	9.4 ± 0.9
PMT + PGE$_2$	(8)	9.7 ± 1.8	10.5 ± 2.0

[a] CDP, 50 μmoles/kg i.p. 60 min before PMT; PGE$_2$, 2.8 μmoles/kg i.p.
3.5 min before PMT; PMT, 100 mg/kg i.p. 1.5 min before sacrifice.
[b] Means ± S.E.M.; numbers of rats in parentheses.
[c] P < 0.01 with respect to control values.

2. PGE$_2$ and Harmaline Tremors

In a completely different type of experiment, tremors were induced
with harmaline, and alkaloid of Peganum Harmala. The olivocerebellar
system is activated by this alkaloid with consequent enhanced firing of
the Purkinje cells. Mao et al. (1975) found that harmaline-induced
tremors are associated with increased cerebellar cGMP. Diazepam protects the animals from the tremorogenic effect of the alkaloid and also
completely inhibits the increase in cerebellar cGMP (Mao et al., 1975).
We have investigated whether PGE$_2$ protects against harmaline-induced
tremors and found that PGE$_2$, given i.p. (2.8 μmoles/kg) 2 min before
harmaline, fully protected all the animals against tremors. The increase
of cerebellar cGMP was also completely inhibited (Table III).

D. DISCUSSION

Prostaglandin E$_2$ prevents against convulsions induced by PMT
and against cerebellar tremor evoked by harmaline. It is noteworthy
that in both cases PGE$_2$ completely inhibited the induced increase of
cerebellar cGMP without affecting its basal levels. In this respect
exogenous PGE$_2$ differs from CDP. CDP, like other benzodiazepines
(Mao et al., 1975), reduced the basal values of cerebellar cGMP (Table
II) but not the rate of increase of cGMP concentrations caused by PMT.
However, the maximal cerebellar concentration of this cyclic nucleotide

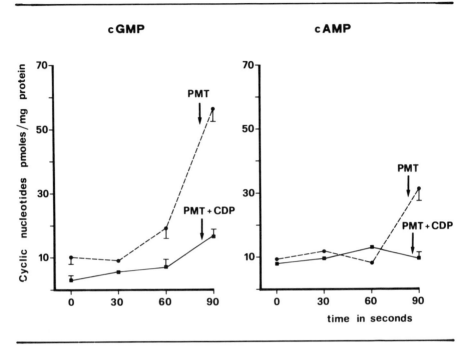

3   Effect of CDP on cerebellar cyclic nucleotides in PMT-treated rats; time course relationship. CDP, 50 μmoles/kg i.p., 60 min before PMT administration. PMT, 100 mg/kg ip., 1.5 min before sacrifice.

after combined treatment with CDP and PMT was lower than 30 pmoles/mg protein, the minimal concentration critical for the onset of convulsions in our experiments. These different effects on cGMP may underline different mechanisms of action for PGE₂ and benzodiazepines. PGE₂, which affects elevated but not basal concentrations of cerebellar cGMP, is likely to have a more specific action.

It is noteworthy that in these experiments the variations of cAMP appear to be dissociable from the convulsions themselves. The consistent increases in cerebellar cAMP occurred only after the onset of convulsions. When increasing doses of PMT were given to rats, or when the effect of a single convulsant dose of PMT was evaluated as a function of time, the increase of cAMP invariably followed that of cGMP. Further support was provided by the use of subconvulsant doses of PMT, which significantly increased cGMP but did not affect cAMP. The enhanced levels of cAMP may possibly be explained as a consequence

Table III. Inhibitory Effect of PGE$_2$ on
Increased Cerebellar GMP
Induced by Harmaline

Treatment[a]	pmoles cGMP/mg protein[b]
Controls	8.6 ± 1.0
PGE$_2$	6.4 ± 0.5
Harmaline	35.6 ± 2.6[c]
PGE$_2$ + harmaline	9.6 ± 1.3

[a] PGE$_2$, 2.8 μmoles/kg i.p. 5 min before sacrifice;
harmaline-HCl: 160 μmoles/kg i.p. 3 min before
sacrifice.
[b] Means ± S.E.M.
[c] Indicates P $<$ 0.001 with respect to control value.

of metabolic events linked to the convulsive state, as suggested also by Sattin (1971).

The selective action of PGE$_2$ against the increased concentrations of cerebellar cGMP can be defined even better by using a different model: harmaline-induced tremors. In this experimental condition, PGE$_2$ completely prevented both the tremors and the increase of cerebellar cGMP without modifying the normal levels of the nucleotide. Harmaline induces tremors by increasing the rhythmic activity of the olivocerebellar pathway (Lamarre et al., 1971), possibly by selective stimulation of climbing fibers (Llinas and Volkind, 1973) which synapse with Purkinje cells, and through collaterals with subcortical cerebellar nuclei (Eccles et al., 1967). Costa and co-workers (1975) first demonstrated the participation of cGMP in the synaptic events of cerebellar neurones and proposed that the cGMP system is, at least in part, concerned with Purkinje cells. Furthermore, in newborn rats, in which the neuronal organization of cerebellar cortex is far from being complete and in which climbing fibers have not yet formed synapses with Purkinje cells, harmaline has neither a tremorogenic action (Henderson and Wolley, 1970) nor the capacity to increase cerebellar concentrations of cGMP (Spano et al., 1975). Additional evidence for a relationship between Purkinje cells and the cGMP system has been provided by Mao et al. (1975) who reported a clearcut decrease of cerebellar cGMP in mice having genetic degeneration of Purkinje cells.

The efficient prevention by PGE$_2$ against the tremors and increase of cerebellar cGMP induced by harmaline indicate that PGE$_2$ may interfere with the synaptic link between climbing fibers and Purkinje cells as has been proposed for diazepam by Mao et al. (1975). Climbing fibers are stimulatory in nature and their activation brings about an enhanced firing rate of Purkinje cells (Eccles et al., 1967).

The increased activity of Purkinje cells would result in a regulation of the spinal motoneurons mediated via GABA release (Obata and Takeda, 1969; Curtis et al., 1970), through subcortical cerebellar nuclei. The identity of the excitatory neurotransmitter released from climbing fibers is not yet clearly established but glutamate is considered to be a possible candidate (Costa et al., 1975; Mao et al., 1974; Young et al., 1974). Since the glutamate-induced increase of cGMP in slices of mouse cerebellum seems to be Ca^{2+}-dependent (Ferrendelli et al., 1974), PGE_2 may well interfere at this level. The findings that prostaglandin dehydrogenase is specifically located in the Purkinje cell layer of the cerebellar cortex (Siggins et al., 1971), and that microelectrophoretic applications of prostaglandins of the E type to these cells interfere with their activity (Hoffer et al., 1969) suggest further that PGE_2 might play a physiological as well as a pharmacological role at this site. The specificity of the prostaglandin-induced prevention of harmaline tremors is strengthened by the lack of effect of PGEs against tremors induced by a different mechanism, such as by oxotremorine (see Horton, 1972). Our investigations have demonstrated for the first time that PGE_2 plays a fundamental role in the regulation of cyclic nucleotides in the CNS. When the possibility that the central effects of intraperitoneally administered PGE_2 are elicited by a peripheral action is considered, it is found to contrast with the observations that other convulsive agents (e.g., picrotoxin) and other tremorogenic drugs (e.g., oxotremorine) are not antagonized by PGE_2 given by the same route at the same doses (Horton, 1972).

E. SUMMARY

Prostaglandin E_2 (PGE_2) exerts a selective action against tremors and convulsions. In the rat, this compound prevents harmaline-induced tremors and is active against PMT-evoked convulsions. PMT convulsions enhance the cerebellar content of both cAMP and cGMP, but the increase of the former nucleotide invariably follows that of the latter and occurs after the onset of convulsions. Subconvulsant doses of PMT significantly increase the basal concentrations of cGMP. PGE_2 has been compared with CDP, an anticonvulsant benzodiazepine, on PMT-evoked seizures. Both compounds prevented convulsions, but affected the cerebellar content of cGMP in different ways. CDP, but not PGE_2, lowered the basal concentrations of cGMP; PGE_2, but not CDP, completely prevented any increase of cGMP elicited by convulsant doses of PMT.

Harmaline elicits tremorgenic activity by its specific activation of the olivocerebellar pathway, which is reflected by a concomitant increase of cerebellar cGMP. PGE$_2$ protected the animals against both effects of harmaline.

Our results provide biochemical bases for the selective anticonvulsant and antitremorogenic actions of exogeneously administered PGE$_2$.

F. REFERENCES

Avanzino, G. L., Bradley, P. B., and Wolstencroft, J. H. (1966). Actions of prostaglandins E$_1$, E$_2$ and F$_{2\alpha}$ on brain stem neurones. *Brit. J. Pharmacol.* 27, 157–163.

Berti, F., Trabucchi, M., Bernareggi, V., and Fumagalli, R. (1973). Prostaglandins on cyclic AMP formation in cerebral cortex of different mammalian species. *Adv. Biosci.* 9, 475–480.

Bradley, P. B., Samuels, G. M. R., and Shaw, J. E. (1969) Correlation of prostaglandin release from the cerebral cortex of cats with electro-corticogram, following stimulation of the reticular formation. *Brit. J. Pharmacol.* 37, 151–157.

Coceani, F., Puglisi, L., and Lavers, B. (1971). Prostaglandins and neuronal activity in spinal cord and cuneate nucleus. *Ann. N. Y. Acad. Sci. U. S. A.* 180, 289–301.

Coceani, F., and Wolfe, L. S. (1965). Prostaglandins in brain and the release of prostaglandin-like compounds from the cat cerebellar cortex. *Can. J. Physiol. Pharmacol.* 43, 445–450.

Costa, E., Guidotti, A., and Mao, C. C. (1975). In: *Proceedings of IXth Congress of Collegium Internationale Neuropsychopharmacologicum,* Paris, in press.

Curtis, D. R., Duggan, A. W., and Felix, D. (1970). GABA and inhibition of Deiters' neurons. *Brain Res.* 23, 117–120.

Duru, S., and Türker, R. K. (1969). Effect of prostaglandin E$_1$ on the strychnine induced convulsion in the mouse. *Experientia* 25, 275.

Eccles, J. C., Ito, M., and Szentagothai, J. (1967). *The Cerebellum as a Neuronal Machine,* Springer-Verlag, Berlin, 335 pp.

Ferrendelli, J. A., Chang, M. M., and Kinscherf, D. A. (1974). Elevation of cycylic GMP levels in central nervous system by excitatory and inhibitory amino acids. *J. Neurochem.* 22, 535–540.

Gilman, A. (1970). A protein binding assay for adenosine 3′, 5′-cyclic monophosphate. *Proc. Nat. Acad. Sci. U.S.A.* 67, 305–309.

Goldberg, N. D., Haddox, M. K., Hartle, D. K., and Hadden, J. W. (1972). The biological role of cyclic 3′, 5′-guanosine monophosphate. In: *Pharmacology and the Future of Man, Proc. 5th Int. Congr. Pharmacology, San Francisco, Cellular Mechanisms,* Vol. 5, pp. 146–169. Karger Publ. Co.

Guidotti, A., Cheney, D. L., Trabucchi, M., Doteuchi, M., Wang, C., and Hawkins, R. A. (1974). Focussed microwave radiation: A technique to minimize post-mortem changes of cyclic nucleotides, DOPA and choline, and to preserve brain morphology. *Neuropharmacology* 13, 1115–1122.

Henderson, G. L., and Wolley, D. E. (1970). Ontogenesis of drug induced tremor in the rat. *J. Pharmacol. Exp. Therapeut.* 175, 113–120.

Hoffer, B. J., Siggins, G. R., and Bloom, F. E. (1969). Prostaglandins E_1 and E_2 antagonize norepinephrine effects on cerebellar Purkinje cells: Microelectrophoretic study. *Science* 166, 1418–1420.

Holmes, S. W., and Horton, E. W. (1968). Prostaglandins and the central nervous system. In: *Prostaglandin Symposium of the Worcester Foundation for Exp. Biol.* (P. W. Ramwell, and J. E. Shaw, eds.), pp. 21–38, Interscience, New York.

Horton, E. W. (1972). *Prostaglandins*, Springer-Verlag, Berlin, Heidelberg, and New York, 197 pp.

Kataoka, K., Ramwell, P. W., and Jessup, S. (1967). Prostaglandins: Localization in subcellular particles of rat cerebral cortex. *Science* 157, 1187–1189.

Kuehl, F. A., Jr. (1974). Prostaglandins, cyclic nucleotides and cell function. *Prostaglandins* 5, 325–340.

Lamarre, Y., De Montigny, D., Dumont, M., and Weiss, M. (1971). Harmaline induced rhythmic activity of cerebellar and lower brain stem neurons. *Brain Res.* 32, 246–250.

Llinas, R., and Volkind, R. A. (1973). The olivo-cerebellar system: Functional properties as revealed by harmaline induced tremor. *Exp. Brain Res.* 18, 69–87.

Lowry, O. H., Rosebrough, N. J., Farr, A. L., and Randall, R. J. (1951). Protein measurement with the Folin reagent. *J. Biol. Chem.* 193, 265–275.

Mao, C. C., and Guidotti, A. (1974). Simultaneous isolation of adenosine-3', 5'-cyclic monophosphate (cAMP) and guanosine-3', 5'-cyclic monophosphate (cGMP) in small tissue samples. *Analyt. Biochem.* 59, 63–68.

Mao, C. C., Guidotti, A., and Costa, E. (1974). The regulation of cyclic guanosine monophosphate in rat cerebellum: Possible involvement of putative amino acid neurotransmitter. *Brain Res.* 79, 510–514.

Mao, C. C., Guidotti, A., and Costa, E. (1975). Inhibition by diazepam of the tremor and the increase of cerebellar cGMP content elicited by harmaline. *Brain Res.* 83, 516–519.

Nicosia, S., and Galli, G. (1975). A mass fragmentographic method for the quantitative evaluation of brain prostaglandin biosynthesis. *Prostaglandins* 9, 397–403.

Obata, K., and Takeda, K. (1969). Release of γ-aminobutyric acid into the fourth ventricle induced by stimulation of the cat's cerebellum. *J. Neurochem.* 16, 1043–1047.

Ramwell, P. W., and Shaw, J. E. (1966). Spontaneous and evoked release of prostaglandins from cerebral cortex of anesthetized cats. *Amer. J. Physiol.* 211, 125–134.

Ramwell, P. W., and Shaw, J. E. (1970). Biological significance of the prostaglandins. *Recent Progr. Horm. Res.* 26, 139–187.

Sattin, A. (1971). Increase in the content of adenosine 3′, 5′-monophosphate mouse forebrain during seizures and prevention of the increase by methylxanthines. *J. Neurochem.* 18, 1087–1096.

Shaw, J. E., Jessup, S. J., and Ramwell, P. W. (1972). Prostaglandin-adenyl cyclase relationships. In: *Advances in Cyclic Nucleotide Research* (P. Greengard and G. A. Robison, eds.), Vol. 1, pp. 479–491, Raven Press, New York.

Siggins, G. R., Hoffer, B. J., and Bloom, F. E. (1971). Prostaglandin-norepinephrine interactions in brain: Microelectrophoretic and histo-chemical correlates. *Ann. N. Y. Acad. Sci. U.S.A.* 180, 302–323.

Spano, P. F., Kumakura, K., Govoni, S., and Trabucchi, M. (1975). Post-natal development and regulation of cerebellar cyclic guanosine mono-phosphate system. *Pharmacol. Res. Commun.* 7, 223–237.

Steiner, A. L., Wehmann, R. E., Parker, C. W., and Kipnis, D. M. (1972). Radioimmunoassay for the measurement of cyclic nucleotides. In: *Advances in Cyclic Nucleotide Research* (P. Greengard and A. G. Robison, eds.), Vol. 2, pp. 51–61, Raven Press, New York.

Uzunov, P., and Weiss, B. (1971). Effects of phenothiazine tranquilizers on the cyclic 3′, 5′-adenosine monophosphate system of rat brain. *Neuropharmacology* 10, 697–708.

Wolfe, L. S., Pappius, H. M., and Marion J. (1976). The biosynthesis of prostaglandins by brain tissue in vitro. In: *Advances in Prostaglandin and Thromboxane Research, Proc. Int. Conference on Prostaglandins* (B. Samuelsson and R. Paoletti, eds.), Raven Press, New York, Vol. I, pp. 345–355.

Young, A. B., Oster-Granite, M. L., Herndon, R. M., and Snyder, S. H. (1974). Glutamic acid: Selective depletion by viral induced granule cell loss in hamster cerebellum. *Brain Res.* 73, 1–13.

Recognins and Their Chemoreciprocals[1]

SAMUEL BOGOCH

*Foundation for Research on
the Nervous System
Boston University
School of Medicine* and
*The Dreyfus Medical Foundation
New York*

[1]The technical assistance of G. Korsh, C. Gramm, and G. Kormby; the cooperation in the provision of serum samples by Dr. M. D. Walker, National Cancer Research Center, National Cancer Institute, Dr. Eli Goldensohn, Department of Neurology, Columbia-Presbyterian Medical Center, and Dr. Charles Fager, Department of Neurosurgery, Leahy Clinic, Boston; the cooperation in the provision of tumor and brain tissue specimens by Dr. William H. Sweet, Dr. Paul Kornblith, and Dr. R. G. DeLong, Departments of Neurosurgery and Neurology, Massachusetts General Hospital, and Dr. Charles Fager, Leahy Clinic; and the provision of immunofluorescent microscopic examination of tumor specimens by Dr. F. H. Hochberg, Department of Neuropathology, Massachusetts General Hospital, are all gratefully acknowledged. The studies reported in this manuscript were not presented at the Satellite Meeting in Madrid in September, 1975 because the author was unable to attend, but are included in this volume at the kind invitation of the editors. The studies reported here on the Recognins and their Chemoreciprocals were supported entirely by Brain Research, Inc. The new products and processes in this communication are the subjects of filed patent applications.

A. INTRODUCTION

For 20 years we have gathered experimental evidence which has consistently supported the proposition that the fundamental chemistry of behavior in man has its roots in the behavior of single cell organisms, indeed of subcellular molecular clusters. When a bacterium meets another bacterium coming through the broth, much of the behavior of which each single cell is capable is related to the mucoid compositions of its outer membranes. Thus, as examples, agglutination, motility, the release of toxins, and sexual mating may be correlated with the specific glycoproteins presented by each cell to its environment. I have proposed that the same rules apply to the formation of intercell contacts in the nervous system, and thus to the formation of stable circuitry, and for experiential modification of this circuitry in training, learning, and memory (the "sign-post" theory; Bogoch, 1965, 1968).

In one attempt to test this hypothesis I considered the state in the brain which is antithetical to the highly ordered one required for survival, one in which cell division is not limited, one in which cytoarchitectural cell–cell relationships are not preserved, one in which recognition functions which normally act to limit further cell division no longer appear to operate between cells. This state is cancer. The brain glial cancer cell is a "dumb" cell, the antithesis of the high information state of normal brain cells, in the sense that cell division proceeds regardless of the fact that space has been used up, and proceeds to the death of the host, and hence of the cancer cells themselves.

In the malignant glial cell of brain tumors, both *in vivo* and *in vitro* in cell culture fermentation, the glycoproteins which increase in training and learning states in normal brain were found to be absent. In their place are simplified compounds from which we have produced low molecular weight products. These products we have called astrocytin and malignin, from the *in vivo* and the *in vitro* sources, respectively.

On examination of another type of disordered brain state, this one a genetic error in development known as Reeler disease in mouse in which certain cells do not follow their normal migration to particular areas in the cortex, we identified another abnormal product replacing that found in the normal state of information in brain.

With the production of these substances in different disorders of recognition, I have proposed the term recognins for those products produced from cells or tissues whose *in situ* precursors appear to have recognition functions in states of order and disorder. At least some of the specificity of the *in situ* precursors is maintained in the recognin products, despite their relatively small molecular weights. This permits the production of antibody-like compounds from serum and other body fluids, and the use of these compounds, here called chemoreciprocals, to provide specific delivery systems to the *in situ* recognin precursors. The first yields of the study of recognins, and their chemoreciprocals have been the development of both an *in vitro* serum diagnostic test and a tissue diagnostic test for malignant brain tumors. The possibilities presented for specific therapy are now being explored. The relevance of these studies is also clear to the understanding of the behavior of cells in normal development, and to the chemistry of tissue transplantation.

For once, the progress in understanding the more complex behavioral systems, those in learning brain, has led to the increased understanding of the chemistry of the behavior of simpler biological systems.

B. SOME METHODOLOGICAL PROBLEMS IN STUDYING THE NEUROCHEMISTRY OF BEHAVIOR OF VERTEBRATES

The nervous system is the most complex biological structure known, and the complexity of its function is attested to throughout the ages by all who have attempted to study it. The attempt to understand the biochemistry of learning and memory in an experiment, therefore, presents a formidable challenge. The variables that require control can be divided into biochemical and behavioral categories. Some of these variables have been discussed by almost all workers in the field, but it is rare that many of them are kept track of in any one experiment. Further, it is likely from the nature of the subject and the unknowns remaining in it at this time, that many of the variables which need attention are as yet unknown.

Some of the behavioral variables are first discussed. It is only in the last decade that most neurochemists have begun to pay attention to the proposition that the state of activity of the brain must be noted at the time of biochemical measurement. Thus a brain that is asleep is clearly not the same brain as one which is fully awake but at rest, or as one which is awake but active.

The activities composed of running, swimming, flapping of wings, climbing, and other vigorous motor pursuits are different from activities of problem solving intensity accompanied by motor quiescence. Both motor activity and problem solving activity are different from the active state of the brain in states of great motivation and in highly emotional states, especially those in which there is fear.

Further, those states broadly referred to as sleep, rest, motor active, problem solving active, motivated active, emotional active (a) have many gradations and (b) usually are each present to different degrees not only in each experiment, but varying throughout the course of longer experiments. For example what may be called the "swim for your life" type of experiment in which a rat must traverse a stretch of very cold water cannot possibly be thought of in the same light as one in which an apparently happy, well-fed, free to move pigeon plays a game with a few buttons, at its own pace. Similarly the "jump or you're going to get it" type of experiment in which the mouse stands on a grid which is periodically electrified to produce a painful and frightening stimulus and usually piloerection and defecation is not the same type of experiment as one in which the mouse may leisurely solve a maze problem for a sure reward at the end, with success at solving, or time to solve the key dependent variable.

It may appear excessive to belabor these points, but even the current literature is full of lofty discussions which compare seemingly disparate biochemical results from two different experiments without noting that they are as different from each other as are those in the above examples. The influence of strong sensory stimuli in themselves are important variables.

Another important variable which is still being ignored is that of time. It is foolish to seek for the biochemical equivalent or correlate of a brain process which occurs as quickly as in the committing to memory of a telephone number in the process of axoplasmic flow whose time constant is hours. The results of one of our studies (Bogoch, 1973) emphasizes this point. That is, the incorporation of ^{14}C-glucose into brain glycoproteins in the resting pigeon is different from that in the training pigeon, but the pattern of incorporation after 10 min of training is different from each of those at 20 min, 30 min, and 1 hr of training.

On the structural and biochemical side, in addition to the usual variables of solubility, *in situ* state of components as compared to final isolated products, the effect of changes in blood flow on localized areas of the brain when small brain areas are being compared, measuring concentrations of enough different related and unrelated substances in each experiment, and problems of compartmentation and pool size in radioactive labeling experiments are only a few examples in this area.

Almost all of these methodological problems are soluble by careful attention to detail, careful observation by the experimenter of what is actually happening to the animal, and control of as many variables as possible, or failing that, by comparison of experiments with different groups of variables. Because of the specialized training which most experimenters receive, close collaboration between two people of neurochemical and experimental psychological disciplines is probably best at this time.

C. PRIMARY AND SECONDARY BIOCHEMICAL PROCESSES IN LEARNING AND MEMORY

An important distinction must be made between energy-generating or membrane-maintaining processes in the nervous system, and the identification of particular molecules and processes which actually mediate the transmission of bioelectrical signal or facilitate storage in a nerve net. The required mechanisms for processing information in the nervous system have been increasingly recognized to involve the following: (1) Sensory input reception. (2) Encoding for transmission: transduction from primary sensory modalities to electrochemical equivalents utilized by nervous system cells. (3) Association and abstraction: association with information previously stored, pattern recognition, abstraction, synthesis of new constructs. (4) Storage: (a) Further encoding, or the same as (2)? (b) Same process for short and long duration? (5) Retrieval: remembering–forgetting. (6) Effector consequences of retrieval: further association and abstraction; discharge into thought, language; motor and affective accompaniments. (7) Supporting chemical reactions for (1) through (6).

1. Sites for These Biochemical Processes (Bogoch, 1968)

The attempt to localize memory, in a biochemical sense, has had certain carryover influences from neuroanatomy and neurohistology. Thus certain efforts concentrate on the notion that memory is mediated by certain distinct brain regions, others on the notion that memory is a totally distributed phenomenon in the brain. Some work focuses on the search for memory traces contained within neurons, or one specific poppulation of neurons; other work concentrates on axonal change, on the growth of new cell processes, on the formation of new cells, on synaptic events, or on glial cells.

2. Some Requirements for Prospective Coding Molecules (Bogoch, 1968)

It is increasingly recognized that information bearing or coding molecules will probably exhibit the following characteristics: sufficient heterogeneity, fixed location, development which correlates with learn-

Table I. Some Biochemical Theories of Memory by Molecular Type

Chemical groups	Changes proposed	Theorist	Ref.	Year proposed
Nucleoproteins	Transformation from random to organized configuration	Halstead	Nakajima and Essman, 1973	1948
Acetylcholine and acetyl-cholinesterase	Increase in quantity and activity, respectively	Rosenzweig et al.	Bogoch, 1968	1958
Nucleic acids or proteins	Self-producing macromolecules	Gerard	Nakajima and Essman, 1973	1959
Ribonucleic acids	Change in base sequences	Hyden	Nakajima and Essman, 1973	1961
	Change in quantity of messenger RNA	Flexner et al.	Nakajima and Essman, 1973	1964
Ribonucleic acids or proteins	Change in base sequences or altered amino acid sequences	Gaito	Nakajima and Essman, 1973	1961
		Ungar	Ungar, 1974	1970
Mucoids (glycoproteins and glycolipids) and glycosyl transferases	Change in concentration and activity respectively; carbohydrate plus protein specificity	Bogoch	Bogoch, 1968	1965, 1968
Catecholamines	Modification of protein effects	Roberts et al.	Nakajima and Essman, 1973	1970
		Essman	Nakajima and Essman, 1973	1973

ing throughout the life of the organism, impairment in structure or concentration or function in correlation with impairment of memory and
learning processes, the ability to demonstrate "recognition" functions,
generative and regenerative properties which correlate with memory
functions, change with certain forms of behavior, and change with
learning.

Some biochemical theories of learning and memory, by molecular
type, are listed in Table I. While it is not possible to discuss these various
theories and the evidence which has accumulated with regard to each, the
references in the table indicate some recent thorough reviews.

In addition to the reviews listed in Table I, the general field of proteins of the nervous system has developed rapidly in the past 10 years
both in terms of definition of structure and metabolism of individual
components. In one volume edited by Lajtha (1970), for example, the
structure of the nerve growth factor proteins is discussed by Shooter and
Varon, axoplasmic flow of nerve proteins by Ochs, the effect of drugs on
protein synthesis in the nervous system by Clouet, inhibition of brain
protein synthesis by Appel, and chemical transfer of learned information
by Ungar. In another volume edited by Davison et al. (1972), brain-
specific proteins are discussed by Moore, by Vincendon et al., by Wa-
recka, and by me; neurotubule and neurofilament proteins by Shelanski
et al., and by Angeletti; glycoproteins in brain by Brunngraber (see
references in Brunngraber to work on their metabolism by Roseman,
DiBenedetta, Dische, Barondes, Irwin, Margolis, Quarles and Brady, For-
man et al., and Boseman et al.); glycoproteins in synaptic membranes
by Breckenridge et al.; and myelin lipids discussed by Mehl, by Folch-Pi,
by Kies, by Eylar, by Mandel, by Morrell et al., and by Marks.

D. HISTORY OF BRAIN GLYCOPROTEIN AND
MUCOID STUDIES

Twenty years ago, the computer analogy to brain function, although
limited, was impressive to some, including me. However, even if this
analogy was cogent, there was no notion of how the billions of cells
in the brain could form such circuitry. How could all these cells grow
together correctly in the first place; and how could these brain circuits
be available to new learning?

Since mucoids (carbohydrates attached to proteins and/or lipids
to form large molecular weight entities) were beginning to be under-
stood to have a key role in determining the chemistry of "recognition"
phenomena in the A, B, O blood group system, and in contact
functions of bacteria, I wondered if they might perform similar functions
in the brain (Bogoch, 1968).

Up to 1963, textbooks of neurochemistry had only a few pages on proteins of the brain. The glycoproteins of the brain were then extracted and found to represent a large and complex fraction of the total brain proteins (Bogoch et al., 1964). The concept that brain glycoproteins are involved in learning processes was developed in studies beginning in 1956 which examined the possibility that the macromolecular carbohydrates in brain have membrane, receptor, and recognition functions. It was demonstrated that a group of substances in the brain with carbohydrate end groups, including the aminoglycolipids and the glycoproteins, have the chemical structural properties which would make them suited to function at membrane surfaces and membrane interfaces (Bogoch, 1968), have the properties of viral receptors (Bogoch et al., 1959; Bogoch, 1960a), have pharmacological specificity (Bogoch and Bogoch, 1959; Bogoch et al., 1962), have molecular specificity related to antigen–antibody reactions (Bogoch, 1960b), occur in high concentration in the brain substance itself (Bogoch, 1962), occur in high concentration in membranes and synaptosome fractions from brain (Bogoch, 1968), show changes in concentration with increasing developmental complexity of the nervous system (Bogoch, 1968), are disturbed when there is regression of higher brain functions (Bogoch, 1962, 1968), and show increases in concentration in particular brain glycoprotein fractions in relationship to operant conditioning training of pigeons (Bogoch, 1965, 1968).

In 1965, when we first demonstrated that the glycoproteins of brain change in concentration during training, I proposed that the glycoproteins and related substances of the brain are involved in information, contact, and communication functions in the nervous system (Bogoch, 1968). The 10 years of work preceding that study and three subsequent years of work on the subject were summarized in the 1968 monograph: *The Biochemistry of Memory: With an Inquiry into the Function of the Brain Mucoids* and the "sign-post" theory was proposed (Bogoch, 1965, 1968). The fine structure of the carbohydrate and amino acid chains of the glycoproteins which appear to be most involved in the brain of the training pigeon were studied in detail (Bogoch, 1970). The function of glycoproteins in the formation of brain circuitry was discussed (Bogoch, 1965, 1969a,b, 1970), and the role of the brain glycoproteins in cell recognition was detailed in further studies on glycoproteins and brain tumors, where there is a regression of these postulated functions for brain glycoproteins (Bogoch, 1972a).

Further to the study of brain glycoproteins in terms of intercell recognition, studies on the brain glycoproteins in the functionally regressive cerebral states of Tay-Sachs' disease have suggested that here there may be a disturbance of intraneuronal recognition (Bogoch 1972b). In the functionally regressive cerebral states observed in the psychoses, our earliest observations on nervous system glycoproteins can be seen from the same perspective. In the case of human cerebrospinal fluid, we found the glycoproteins to be present in surprisingly high concentration, to increase with maturation, to show certain deficiencies in psychotic illness, and to return toward normal with recovery (see Campbell et al., 1967). Thus if the brain glycoproteins are indeed involved in establishing functionally correct brain circuitry then disorders of brain glycoproteins may well be responsible for the functional disorders of thought, mood, and behavior.

E. THE "SIGN-POST" THEORY OF BRAIN CIRCUITRY

I have proposed that the glycoproteins of the nervous system are involved in cell recognition, contact, annd position functions and that these properties are responsible for the stable intercell circuitry necessary for information handling and storage by the nervous system (Bogoch, 1965, 1968). That is, the glycoproteins determine which nerve cells grow together in contact during development to permit transmission as well as the facility of transmission at a given synaptic junction throughout life.

"A single carbohydrate end group might determine whether or not transmission of an impulse is facilitated between neuron A and B. Consider that in the presynaptic membrane belonging to neuron A there is a glycoprotein, glycoprotein A^1, whose carbohydrate end group can be two types, which determine whether contact is or not taking place with the postsynaptic membrane belonging to neuron B. Thus, for example, glycoprotein A^1 may exist with its end group a galactosamine residue, but with the postsynaptic membrane B possessing the specific receptor B^1 (possibly, but not necessarily, also a glycoprotein) specific only for a galactose residue. In the "resting" state A^1 does not combine with B^1 and no synaptic contact results. Thus, although transmission is possible between A and B, it may not occur without facilitation. To state it another way, the threshold for transmission between neurons A and B may be an inverse function of the amount of physical contact between the membranes of A and B; the more contact, the lower the threshold.

"A synthetic mechanism, a galactose transferase here taken as only one example, could be present at or in A, which could on the proper stimulus attach a galactose residue to the galactosamine end group of A^1, thus changing the galactosamine end group to a galactose end group for A^1. This galactose end group of A^1 would combine immediately with receptor B^1, ionically or covalently, the contact between A and B would be increased, the threshold lowered, and transmission between A and B facilitated. It is possible that the "proper stimulus" for attachment of galactose is the passage of an excitatory impulse of sufficient intensity through A. The question of whether A can fire B will depend on (1) whether the synthetic reaction necessary to attach the correct end group on A^1 is at hand and immediately available, or if repressed, is able to be derepressed; and (2) whether the receptor B^1 is at hand to react with A^1 galactose. These paired configurations might be laid down with complete or relative specificity by genetic coding mechanisms and realized in the morphogenesis of the nervous system. Thus the chemical specificities of A and B which allow them to grow together in synaptic contact in the first place would be the same specificities determining their contact and the transmission of an impulse between them throughout the life of the organism.

"The influence of experience would enter in terms of (1) the frequency of stimuli, or both, passing through A causing it to fire B. That is, the potential to synthesize A^1-galactose may require the activation by strong and/or repeated stimulation. (2) The competitive pathways available. That is, neurone A could transmit to neuron C, A to D, etc. All might be programmed genetically as possibilities, then a combined selection-instruction mechanism brought to bear by experience to determine which pathway is selected, or which alternative pathways are preferred, and their order of preference. A DNA-specified repression–derepression type of induction requiring experiential input for activation could well be involved for the A^1-galactose transferase reaction. The potential of the system may be quite extensive, as genetically programmed, but each component would require experiential realization in order to perform. That is, the degree of derepression in each instance would be a direct function of experiential input.

"For A^1-galactose, it might be necessary to substitute A^1- (glycosaminoglycan)$_n$-galactose, "n" representing the number of residues actually required to bridge the synaptic cleft so that the molecule A^1 may reach B^1.

"Even greater complexity and selectivity could be achieved by having more than one specific glycoprotein and receptor present per synapse, thus A^1, A^2, A^3, A^n and B^1, B^2, B^3, and B^n.

"Chemical coding" of information in the brain would thus be visualized as a specific set of instructions determining whether, and under what conditions, each neuron will fire any other accessible neurone. The actual information codes in glycoprotein A^1 could be no more than that required to define whether or not B^1 would be contacted. The mucoids would thus act as switching mechanisms, "sign-posts" which route transmission. The mucoids would thus be the chemical basis of the make-break mechanisms of the brain's circuitry, the chemical basis of the establishment of the cluster of specific circuits which constitute a memory trace. They might also underlie specificities of contact between glia and neurons" (see Bogoch, 1968).

The experimental evidence in support of this "sign-post" theory, summarized in the above references, has included the fact that a change in the structure and function of the brain glycoproteins should accompany both the expression of, and the regression of, these higher nervous system functions.

The general field of glycoproteins outside the nervous system has grown in the last few years and has provided evidence from nonbrain systems relevant to the postulated or demonstrated functions of brain glycoproteins. For example, lectins and cell-agglutinating and sugar-specific proteins are useful in the study of the native glycoproteins of the cell surfaces and of the basic recognition and contact functions of cells. The properties of glycoproteins on the cell surface are seen to be disturbed in some cancer cells, are thought to be related to transplantation antigens, have been shown to be cell surface receptors for viruses and endotoxins, and are involved in the attachment and penetration of the ovum by the spermatozoan (see references in Bogoch, 1973).

These various recent confirmations of the role of glycoproteins in cell surface recognition phenomena outside of the nervous system are clearly relevant to the evidence and hypothesis on glycoproteins of the nervous system.

F. FURTHER EVIDENCE IN SUPPORT OF THE BRAIN GLYCOPROTEIN "SIGN POST" THEORY IN LEARNING AND MEMORY

1. Changes in Glycoprotein 11A in Relation to the "Amount of Learning"

Although many experimental protocols have been published for the study of the "all-or-none" acquisition of information in a variety of experimental animals, it is a more difficult task to express a graded response in terms of the amount learned in a given time by an individual subject. In order to approach this problem in the training pigeon, as it might relate to brain glycoproteins, we have done the following experiment.

The pigeons were male White Carneaux. They were maintained at approximately 85% of their free-feeding body weights. The apparatus used was a two-key operant chamber for pigeons according to Ferster and Skinner. First, all birds were trained to peck equally at the two response keys in order to obtain food reinforcement. After this preliminary training a concurrent VI I VI I schedule was instituted. On this schedule, responding on one key occasionally (on the average of once each minute) resulted in a delay interval of 20 sec that was followed by food. Responding on the other key resulted only in the delay (again, on the average of once each minute). During the 20-sec delay, the experimental chamber was completely dark, and responses were ineffective. Thus, responding on only one of the keys is effective in producing food. The subjects were run for 21 sessions, each session terminating after 30 reinforcements were received. Following the last session, all subjects were sacrificed by dropping the pigeon into a dry ice-acetone bath. Each pigeon brain was coded and extracted individually for brain glycoproteins, and 16 groups separated on column chromatography with Cellex D as previously described (Bogoch, 1965, 1968). The chemical analyses and Folin–Lowry quantitative determinations of protein were done "blind" with regard to the performance of each pigeon.

The degree to which the responses deviated from a random 50–50 response to the two keys represented the preference for the correct key. The results of this deviation from random are expressed as deviations from 50, the highest deviation being 50 and the lowest being zero. The pigeons were ranked according to their scores and then divided into three groups: the highest group scoring in the range of 28–50, the middle group scoring in the range of 10–24, and the lowest group scoring in the range of 0–9. Because of the small numbers of subjects in each group, only a trend can be perceived. The pigeons which demonstrated the highest "amount of learning" had the highest mean amount of brain glycoprotein 11A, 2.63 mg/g per gram of wet weight of brain tissue; those of the middle group had 0.65 mg/g, and those of the lowest group had 0.56 mg/g. The concentration of brain glycoprotein 10B did not show any relationship to rank order. Since these were long-term results observed after 21 sessions of training, these results were in confirmation of the earlier results on the changes in brain glycoprotein 10B which was elevated early, then normal later (Bogoch, 1965, 1968).

2. Carbohydrate Constituents of Pigeon Brain Glycoproteins 10B and 11A

In earlier studies it was shown that the carbohydrate constituents of glycoproteins 11A and 10B (10B has two separate fractions, 10B, and 10B$_2$, in the pigeon) are quite different from each other, although their amino acid chains appear quite close in composition (Bogoch 1965, 1968). Thus, a change from 10B to 11A represents a major structural change in terminal carbohydrate units, which would correlate with the notion that specific kinds of connections are being favored o that entirely new kinds of connections are being formed. Thus, 11A has considerably less hexose and hexosamine than does either $10B_1$ or $10B_2$. These differences are sufficient that specific antibodies have been made to these substances which distinguish between them. This type of specificity is therefore of a sufficiently high order that it might account for specificity of contact between neuronal membranes of two different neurons.

3. Incorporation of ^{14}C-Glucose into Training and Resting Pigeon Brain Glycoproteins

Glucose labeled in the first carbon was injected intravenously into pigeons at rest, or before they engaged in a training procedure for varying periods of time before sacrifice in a dry ice–acetone bath. The brain glycoproteins were extracted and separated as previously described (Bogoch, 1965, 1968). At rest for 30 min only slight incorporation was observed in groups 2, 3, and 11B. Much more incorporation was observed at only 10 min of training. Furthermore, the nature of the most actively labeled groups was a function of time. Thus, at 10 min training, groups 1, 10A, and $10B_1$ and $10B_2$, 11B, 12, and 13 were most active. At 60 min, groups 1 and 2, that are associated with the lateral dendritic processes (Bogoch, 1968), and 10B proteins that are associated with astrocytic glia were most active. This change with time of the groups that were most actively labeled indicated that the sequential activity of glycoproteins of different cellular organelles as well as of different cell types (glia and neurons) will require further individual examination.

What is clear is that there is an extremely active turnover of glycosidically bound glucose in the brain glycoproteins of training pigeons and that much of this activity occurs within minutes. We have previously noted (Bogoch, 1965) that these changes occurred earlier than at those times usually examined in such studies, and the present data confirm the importance of examining discrete and short time periods in relation where possible to specific glycoproteins and specific cell fractions.

While glucosamine was found in one study to be slowly incorporated during training (Routtenberg et al., 1971), confirmation of our finding of early and rapid incorporation of sugars into brain glycoproteins has been obtained for training mice when tritiated fucose was the injected sugar (Damstra-Entingh et al., 1974; Routtenberg et al., 1974). In further confirmation of both early incorporation (30 min) and the relevance of the functional state, "aggregated" mice were shown to incorporate more radioactivity of subcutaneously injected ^{14}C-glucose and ^{14}C-mannose into their brains than did "isolated" mice (DeFeudis, 1972, 1973). In addition, recent work with pigeons has confirmed many aspects of our isolation procedures for brain glycoproteins and the incorporation patterns of glucose (in this case, 2-³H-glucose) with training (Irwin and Barraco, 1975).

4. Glycoproteins Are in High Concentration at the Synapse

With the development of the new stains for the electron microscope which visualize large molecular weight carbohydrate materials, the periodic acid stains, it has been possible to demonstrate visually the presence of glycoproteins in high concentration at the synaptic cleft. Periodic acid-silver methenamine shows stained material separating nerve and glial processes from each other in the neuropil. Staining of the intercellular space is sharply increased in the region of the synaptic cleft. With phosphotungstic acid, as with periodic acid-silver methenamine, stained layer outlines cell and nerve processes. Staining of the intercellular space is enhanced in the synaptic cleft (Rambourg and Le Blond, 1969).

G. INVERSE RELATIONSHIP OF MUCOID-CONTACT FUNCTIONS AND DNA REPLICATION: THE REGRESSIVE STATE OF BRAIN TUMORS

The search for the chemical expression of higher information functions of brain cells has led to findings of structural changes in similar molecules on regression of these higher functions in brain tumors.

The growth of ependymoma-glioma tumors subcutaneously in the mouse provided the opportunity to study a brain tumor growing outside the nervous system, and at the same time to demonstrate that the glycoproteins of the essentially normal brain in the same animal are influenced by a diffusible distance factor from the subcutaneously growing brain tumor (Bogoch, 1972a). This factor may be responsible for the induction of a precancerous or early cancerous change in normal cells.

The concentration of total protein-bound hexose in brain under the influence of distant tumor is in agreement with the data obtained for human brain tumors. It is also in general agreement with the previous observations that the concentration of protein-bound hexose in brain is greater: (1) the more complex the anatomic structure of the brain, (2) the more ordered the chemical structure, as in membranes opposed to cytoplasmic cell constituents, and (3) the more complex or active the functional state of the animal (Bogoch, 1965, 1968). Thus the greater the "experiential" or environmental informational content or activity of a given structure of function, the greater the concentration of protein-bound hexose. The training brain cell, with relatively high experiential informational content, contains more protein-bound hexose (Bogoch, 1965, 1968). The tumor cell, with relatively low experiential informational content, but high "genetic" (mitotic) activity, contains less protein-bound hexose.

This relationship has been generalized to the proposition I made that when DNA and cell replication functions are stimulated, as in tumors, normal mucoid biosynthesis is inhibited (Bogoch, 1972a). During normal morphogenesis of brain, DNA and cell division would be more active at one stage, and mucoid biosynthesis for specific inter-neural connections would be more active at another stage.

This postulated inverse relationship of mucoid-contact functions to DNA replication could account for the cessation of DNA synthesis in retinal ganglion cells observed to be correlated with the time of specification of their central connections (see Bogoch, 1972a). The theory would also be consistent with the reported relationship of malignancy to loss of contact inhibition (see Bogoch, 1972a) if contact recognition is indeed a function of mucoids. This cell surface difference in malignancy has been related to the presence of an agglutinin with specificity for the N-acetylglucosamine determinant (see Bogoch, 1972a). If nucleic acid bases, uridine and cytidine, were available either for nucleic acid synthesis or for transferase activity for mucoid synthesis, but not for both simultaneously, then a control mechanism would be at hand for determining which of the two cell functions—replication or contact positioning—occurred.

H. ASTROCYTIN AND MALIGNIN: TWO ANTIGEN-LIKE FRAGMENTS (RECOGNINS) RELEVANT TO MALIGNANT BRAIN GLIOMAS

Ten years ago we found that glycoprotein fraction 10B was increased in gliomas compared to tissue from normals and from patients afflicted with other disorders (Bogoch and Belval, 1966). The increase was in the protein components but the carbohydrate components were reduced (Bogoch, 1972a). There was an increase in 10B in Tay-Sachs' disease also, but here the increase was seen to be in the carbohydrate components as well as the protein (Bogoch and Belval, 1966). Antisera prepared to crude fraction 10B from Tay-Sachs' brain, on immuno-fluorescent study of brain sections containing tumor cells, did not stain the tumor glia cells, did not stain normal resting glia, nor blood vessels, nor neurons, but stained only reactive glia (Benda et al., 1970). Now, in retrospect, in view of the data I present, it is clear that this Tay-Sachs' antibody was an antibody to a recognition substance (recognin), a component of reactive glia which are present in high concentration in both Tay-Sachs' disease and brain tumors. I have now found that the normal brain cell recognition substances (recognins) which operate during the high information states of memory and learning have been replaced by another recognition substance in the tumor cell itself. The identification of this recognition substance has led to the following (Bogoch, 1976a,b):

1. The production of the first cancer antigen-like compound of known structure, astrocytin, from crude fraction 10B of brain tumors.

2. The synthesis of new compounds (Target reagents) which contain the immunochemical specificity of the *in situ* cancer antigen.

3. The use of these target reagents has permitted the isolation of pure specific antibody-like materials (Target-attaching globulins, TAG) from human and animal serum.

4. The quantitative measure of serum TAG has provided the first quantitative serum diagnostic method for malignant brain tumors.

5. The TAG compounds react with the cancer antigen-like compounds in quantitative precipitin, Ouchterlony double diffusion reactions, and label *in situ* glial cancer cells specifically in double layer immunofluorescence—hence a new tissue diagnostic method for malignant glia as well.

6. The cancer antigen-like material has been produced by glioma cancer cell culture fermentation. This product, malignin, is structurally unique from the astrocytin, but shares its immunochemical properties. Its bulk production permits the mass use of Target reagent in diagnostic testing.

7. The percentage of extractable cancer protein which is malignin increases as the malignancy of the cancer growth increases.

8. Computer search of the library of protein structures reveals that astrocytin and malignin have unique structures, and that the only proteins even remotely related are four in number, each a respiratory protein. This finding may be of great significance since it might provide the first biochemical insight into the basis of the respiratory activity of cancer cells and their dominance. It may also provide new therapeutic means for their destruction because of the following point.

9. TAG compounds have been found to be highly cytotoxic to malignant glia *in vitro*.

10. The implications of these findings extend beyond cancer diagnosis and treatment and into tissue and organ specificity, transplantation rejection, specific pharmacological delivery systems, and developmental disorders.

11. The third recognition substance of this type has been identified in Reeler disease in mouse brain to be of abnormal molecular weight.

12. With the productions of this third product, the family name for the recognition substances, recognins, is here proposed, together with the name of chemoreciprocals for their antibody-like counterparts.

I. THE STRUCTURE OF ASTROCYTIN AND MALIGNIN

The two unique small molecular weight protein fragments, named astrocytin and malignin, are produced from *in vivo* brain glial tumor and *in vitro* cancer cell fermentation, respectively. They bear specific tissue antigen and other immunochemical properties characteristic of the malignancy.

Astrocytin has a molecular weight of approximately 10,000 and the following approximate composition: Asp 9, Threo 5, Ser 6, Glu 13, Pro 4, Gly 6, Ala 9, Val 4, 1/2 Cys 2, Meth 1, Isoleu 2, Leu 8, Tyr 2, Phe 3, Lys 8, His 2, Arg 4; total 88.

Malignin, repeatedly produced over 60 generations from glioma cell culture fermentation, has a molecular weight of approximately 10,000 and the following approximate composition: Asp 9, Threo 5, Ser 5, Glu 13, Pro 4, Gly 6, Ala 7, Val 6, 1/2 Cys 1, Meth 2, Isoleu 4, Leu 8, Tyr 3, Phe 3, Lys 6, His 2, Arg 5; total 89. The comparison of these two structures with that of structurally related proteins found by computer search is shown in Table II.

The percent of total extracted cancer cell protein which is malignin increases with the degree of malignancy of the growth (Fig. 1).

J. ANTIBODIES TO ASTROCYTIN AND MALIGNIN

Antibodies prepared in rabbits to astrocytin are specific to both astrocytin and malignin, react with them in quantitative precipitin and Ouchterlony double diffusion reactions, and label *in situ* glial cancer cells specifically in double layer immunofluorescence (hence a new tissue diagnostic method).

Figure 2 shows a section of brain removed at surgery in the case of a malignant glial tumor. The section is stained with the new product TAG (target-attaching globulin), an antibody-like protein fragment prepared from rabbits challenged with the recognin, and with fluorescent goat anti-rabbit serum. In contrast to the results obtained with anti-Tay-Sachs' antisera where only reactive (normal) glia and no tumor glia stained, here we see the opposite, that is, only tumor glia stain.

This new tissue diagnostic test promises to be of use (1) in identifying the presence of a malignant tumor (vs. a benign one) at surgery, and (2) in defining whether there are still malignant glia present in the

Table II. Comparison of the Structures of Astrocytin and Malignin to Nearest Structures[a] by Computer Search

Amino acid	Astrocytin	Malignin	Cytochrome b5[a]	Ferredoxin, leucaene glauca[a]	Ferredoxin, alfalfa[a]	Acyl carrier, E. coli*	Neurophysin, bovine	Neurophysin, pig	Gonadotropin, release
Aspartic acid	9	9	9	10	8	7	2	3	0
Threonine	5	5	6	4	6	6	2	2	0
Serine	6	5	5	7	8	3	6	7	1
Glutamic acid	13	13	14	13	13	14	9	9	0
Proline	4	4	3	5	3	1	8	7	1
Glycine	6	6	6	7	7	4	16	14	2
Alanine	9	7	4	6	9	7	6	7	0
Valine	4	6	4	6	9	7	4	2	0
½ cysteine	2	1	0	5	5	0	14	14	0
Methionine	1	2	1	0	0	1	1	1	0
Isoleucine	2	4	4	4	4	7	2	2	0
Leucine	8	8	7	10	6	5	6	7	1
Tyrosine	2	3	3	3	4	1	1	1	0
Phenylalanine	3	3	3	3	2	2	3	3	0
Lysine	8	6	7	5	5	4	2	2	1
Histidine	2	2	7	1	2	1	0	0	1
Arginine	4	5	3	2	1	2	7	5	1
Asparagine	0	0	0	0	1	1	3	2	0
Tryptophan	0	0	1	1	1	0	0	0	1
Glutamine	0	0	0	4	3	4	5	4	0
Total no. residues	88	89	87	96	97	77	97	92	10
Molecular weight[b]	8000[b]	10,000[b]	10,035	10,588	10,483	8509	10,065	9488	1201

[a] Only four close structures found; other three shown for comparison.
[b] By thin-layer gel chromatography; calculated molecular weights of above structures: astrocytin, 9,690; malignin, 10,067.

214

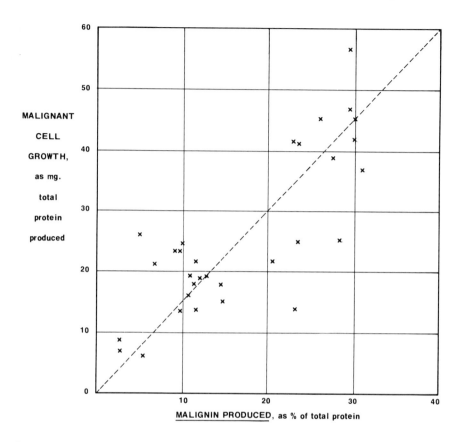

1 Increase in malignin produced, as percent of total protein, with increase in malignant cell growth.

edges of the tissue removed at surgery, hence an indication of whether all of the tumor was removed. Since TAG compounds also are highly cytotoxic to these cancer cells *in vitro*, their therapeutic potential requires exploration in man.

K. SERUM TAG TEST FOR MALIGNANT BRAIN TUMORS

Synthesis of astrocytin or malignin complexes with carriers such as cellulose derivatives produces solid topographic antigen-like reagent templates (Target). These reagents react with antibody-like globulins in human serum and from the resulting complexes, Target-attaching globu-

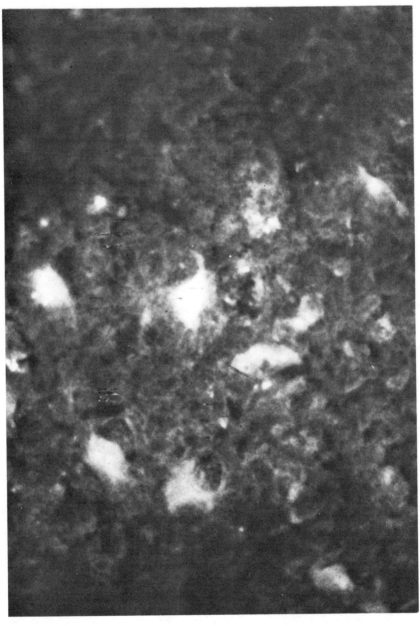

2   Malignant glial cells in human brain tumor stained in immunofluorescence with TAG.

lins (TAG) can be isolated. Figure 3 shows the quantity of serum TAG, in micrograms of protein per milliliter for normals and tumors. TAG is 0 to 130 in 71, greater than 130 in 3 (mean 54.3) in normals and non-tumor medical and surgical disorders; 137 to 650 in 20, less than 130 in 2 (mean 246.4) in malignant brain tumors; 157 to 270 in 5 secondary malignant tumors in brain; 288 and 442 in 2 cancers of the prostate with no apparent brain secondaries; 31 and 36 in 2 sarcomas with no apparent secondaries to brain; and 9 with as yet uncertain clinical diagnoses. The above data represent two successive blind studies and a total of 114 serum specimens, and support the utility of the serum TAG determination as a diagnostic procedure for malignant brain tumors. The serum TAG diagnosis of primary malignant brain tumors in these studies is 91% accurate, of primary plus secondary malignant brain tumors 92.6% accurate. There may be 4% false positives in the combined normal and medical-surgical disorder group. There were no false positives in normal symptom-free individuals. The group of malignant brain tumors is distinguished from the normal and medical-surgical disorder group at a level of $p < 0.0001$.

L.   STRUCTURAL RELATION TO RESPIRATORY PROTEINS

Computer search (Dayhoff, 1972) shows astrocytin and malignin to be unique in structure. The closest structures, surprisingly, are certain respiratory proteins: cytochrome b_5 (human), ferredoxin of leucaene glauca, ferredoxin of alfalfa, and the acyl carrier protein of Escherichia coli (see Table II). (Note especially the anaerobic properties of the ferredoxins; neoplastic cells can survive for long periods of time in the absence of glucose and oxygen, even in the presence of cyanide; Warburg et al., 1958). However while astrocytin and malignin each have only 2 histidines, cytochrome b_5, for example, has 7 histidines, and there are 6 to 13 other structural differences in each of the above structural neighbors. On the other hand, no proteins in the computer library are anywhere near as close as the above four, the lack of even superficial similarity being chiefly due to the highly unusual structural characteristic of astrocytin and malignin, the fact that each has 13 residues of glutamic acid in an 88 and 89 residue total, respectively. If the *in situ* equivalents of the cancer recognins in fact have respiratory functions in cancer cells, since they account for such a large proportion of the extracted cell protein, a biochemical mechanism may now be apparent for the first time to account for the voracious respiratory activity of the cancer cell. This might also explain the absolute increase in malig-

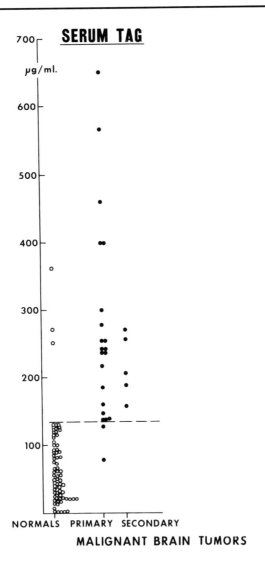

SERUM TAG

NORMALS PRIMARY SECONDARY

MALIGNANT BRAIN TUMORS

3

Serum TAG test for malignant brain tumors.

nin with increased malignancy (Fig. 1), and why TAG is cytotoxic to cancer cells since it binds an essential respiratory protein.

M. DISCUSSION

The definition of recognition substances in mature brain cells which operate during high information states of memory and learning (Bogoch, 1968) has led to the observation that these substances are replaced in cancer by *in situ* compounds which have the immunochemical specificities of astrocytin and malignin. These new compounds characterize the low information (dumb) state of the cancer cell in that cell division proceeds regardless of the fact that the space available has been used up, and proceeds to the death of the host, and hence of the cancer cells themselves. With the production of these first two compounds of the group, the family name of recognins is proposed. For the antibody-like compounds to the recognins, the name chemoreciprocals is proposed.

The recognins have immediate utility in the diagnosis of brain malignancies, and possibly others, and represent a major advance in our understanding of the structural basis of cancer antigens. The same technology is being applied by us to other cancers. The recognins may simultaneously provide new insight into the respiratory survival efficacy of cancer cells and new therapeutic means to destroy them.

These findings are also clearly relevant to other tissue and organ specificity studies, as in transplantation rejection, to disease specificity, and to means of specific delivery of pharmacological agents to particular subcellular and cellular addresses. There are also clear applications to defining and treating developmental disturbances, as those of cell positioning. Thus for example, we have now isolated a third recognin—that in Reeler disease of mouse brain. In this genetic movement disorder in which particular groups of nerve cells fail to migrate to their normal position, Reeler recognin is obviously abnormal, since it has a molecular weight of 3,600 to 5,000, compared to 8,000 in normal mouse brain.

N. CONCLUSION

There is a biochemistry of learning and memory. There are methodological problems in its study. There are a number of theories now being explored experimentally. The "sign-post" theory of the function of brain glycoproteins in the establishment of brain circuitry and its relationship to learning and memory are discussed in some detail (Bogoch, 1975).

Studies to date provide evidence that the "expression" of brain function in learning and memory is related to the proposed higher recognition functions of the brain glycoproteins. Other studies have demonstrated the "regression" of these higher recognition functions in the loss of carbohydrate moieties of the brain glycoproteins occurring in brain tumors. This, in turn, has led to the extension of the "sign-post" concept to the proposal that mucoid biosynthesis is inversely related to DNA replication. That is, DNA and cell division are inhibited during cell positioning and the formation of intercell contacts (expression of the proposed "sign-post" experiential function), and mucoid biosynthesis is inhibited when DNA and cell division are active (regression of "sign-post" function).

Thus, brain glycoproteins may well be involved in the developmental establishment of normal brain circuitry, and in its daily maintenance and change in relation to experiential input; and disorders in brain glycoproteins may be associated with regressive disorders of higher brain functions as occur in brain tumor and psychoses. Many other neurochemical systems may participate in both the primary and the secondary biochemical processes in learning and memory. Astrocytin, malignin, and Reeler recognin are the first three recognins (recognition substances) defined. A serum diagnostic test for malignant brain tumors using synthetic recognins and their chemoreciprocals is described.

O. REFERENCES

Benda, P., Mori, T., and Sweet, W. H. (1970). Immunofluoresent studies of a glial-specific protein (Bogoch 10B). *J. Neurosurg.* 33, 281.

Bogoch, S. (1958). Studies on the structure of brain ganglioside. *Biochem. J.* (*London*) 68, 319.

Bogoch, S. (1960a). Interaction of viruses with native brain substrates. *Neurology* 10, 439.

Bogoch, S. (1960b). Demonstration of serum precipitin to brain ganglioside. *Nature* (*London*) 185, 392.

Bogoch, S. (1962). Aminoglycolipids and glycoproteins of human brain: New methods for their extraction and further study in the sphingolipidoses. In: *Proc. International Conference on the Sphingolipidoses* (S. M. Aronson, and B. W. Volk, eds.), p. 249, Academic Press, New York.

Bogoch, S. (1965). Brain proteins in learning: Findings in the experiments in progress discussed at the NRP work session—June 1965. *Neurosci. Res. Prog. Bull.* 3, 38.

Bogoch, S. (1968). *The Biochemistry of Memory: With an Inquiry into the Function of the Brain Mucoids*, Oxford University Press, New York.

Bogoch, S. (1969a). Nervous system proteins. In: *Handbook of Neurochemistry* (A. Lajtha, ed.), Vol. I, Plenum Press, New York.

Bogoch, S. (1969b). Brain circuitry and its structural basis. In: *Future of the Brain Sciences*, pp. 104–113, Plenum Press, New York.

Bogoch, S. (1970). Glycoproteins of the brain of the training pigeon. In: *Protein Metabolism of the Nervous System* (A. Lajtha, ed.), pp. 535–569. Plenum Press, New York.

Bogoch, S. (1972a). Brain glycoprotein 10B: Further evidence of the "signpost" role of brain glycoproteins in cell recognition, its change in brain tumor, and the presence of a "distant factor." In: *Functional and Structural Proteins of the Nervous System* (A. N. Davison, I. G. Morgan, and P. Mandel, eds.), pp. 39–54, Plenum Press, New York (abstr. 3rd Intern. Meeting Soc. Neurochem., 1971).

Bogoch, S. (1972b). Brain glycoproteins in intercell recognition: Tay-Sachs' disease and intraneuronal recognition. In: *Sphingolipids, Sphingolipidoses and Allied Disorders* (B. W. Volk, and S. M. Aronson, eds.), pp. 127–149, Plenum Press, New York.

Bogoch, S. (1973). Brain glycoproteins and learning: New studies supporting the "sign-post" theory. In: *Current Biochemical Approaches to Learning and Memory.* (W. B. Essman and S. Nakajima, eds.), pp. 147–157 Spectrum Publications, New York.

Bogoch, S. (1975). In: *The Nervous System, Vol. I, The Basic Neurosciences* (D. B. Tower. and R. Brady, eds.), pp. 591–600, Raven Press, New York.

Bogoch, S. (1976a). In: *Recent Advances in Sphingolipidoses and Allied Disorders* (B. W. Volk and L. Schneck, eds.), pp. 555–566, Plenum Press, New York.

Bogoch, S. (1976b). In: *International Conference on the Immunobiology of Cancer,* Nov. 3–4. (*Ann. N. Y. Acad Sci,* in press).

Bogoch. S., and Belval, P. C. (1966). In: *Cerebral Sphingolipidoses* (S. M. Aronson and B. W. Volk, eds.), p. 273, Academic Press, New York.

Bogoch, S., and Bogoch, E. S. (1959). Effect of brain ganglioside on the heart of the clam. *Nature (London).* 183, 63.

Bogoch, S., Lynch, P., and Levine, A. S. (1959). Influence of brain ganglioside upon the neurotoxic effect of influenza virus in mouse brain. *Virology* 7, 161.

Bogoch, S., Paasonen, M. D., and Trendelenburg, U. (1962). Some pharmacological properties of brain ganglioside. *Brit. J. Pharmacol.* 18, 325.

Bogoch, S., Rajam, P. C., and Belval, P. C. (1964). Separation of cerebroproteins of human brain. *Nature (London)* 204, 73.

Campbell, R., Bogoch, S., Scolaro, N. J., and Belval, P. C. (1967). Cerebrospinal fluid glycoproteins in schizophrenia. *Amer. J. Psychiat.* 123, 952.

Damstra-Entingh, T. D., Entingh, D. J., Wilson, J. E., and Glassman, E. (1974). Environmental stimulation and fucose incorporation into brain and liver glycoproteins. *Pharmacol. Biochem. Behavior* 2, 73–78.

Davison, A. N., Mandel, P., and Morgan, J. G. (eds.) (1972). *Functional and Structural Proteins of the Nervous System*, Plenum Press, New York.

Dayhoff, Margaret O. (ed.) (1972). *Atlas of Protein Sequence and Structure*, National Biomedical Research Foundation, P. O. Box 629, Silver Spring, Md. and Supplement 1. (1973).

DeFeudis, F. V. (1972). Effects of isolation and aggregation on the incorporation of carbon atoms of D-mannose and D-glucose into mouse brain. *Biol. Psychiat.* 4, 239.

DeFeudis, F. V. (1973). Effects of *d*-amphetamine on the incorporation of carbon atoms of D-mannose into the brain and sera of differentially housed mice; short-term reversibility of these effects. *Biol. Psychiat.* 7, 3.

Irwin, L. N., and Barraco, R. A. (1975). Abstracts 5th International Meeting of the International Society for Neurochemistry, Barcelona, Sept. 2–6.

Lajtha, A. (ed.) (1970). *Protein Metabolism of the Nervous System*, Plenum Press, New York.

Nakajima, S., and Essman, W. B. (1973). Biochemical studies of learning and memory: An historical overview. In: *Current Biochemical Approaches to Learning and Memory* (W. B. Essman, and S. Nakajima, eds.) Spectrum-Wiley, New York.

Rambourg, A., and Leblond, C. P. (1969). Localisation ultrastructurale et nature du materiel colore au niveau de la surface cellulaire par melange chromique-phosphotungstique. *J. Microsc.* 8, 325.

Routtenberg, A., George, D. R., and Davis, L. G. (1974). *Behav. Biol.* 12, 461.

Routtenberg, A., Holian, O., and Brunngraber, E. G. (1971). Memory consolidation and glucosamine-1-C^{14} incorporation into glycoproteins. *Trans. Amer. Soc. Neurochem.* 2, 103.

Ungar, G. (1974). Molecular coding of information in the brain. In: *Biological Diagnosis of Brain Disorders* (S. Bogoch, ed.), Plenum Press, New York.

Warburg, O., Gawehn, K., Geissler, A. W., Schroder, W., Gervitz, H. S., and Volker, W. (1958). *Arch. Biochem. Biophys.* 78, 573.

12

A Neurobiological Theory of Action of Lithium in the Treatment of Manic-Depressive Psychosis

ARNOLD J. MANDELL
SUZANNE KNAPP

Department of Psychiatry
University of California at San Diego
La Jolla, California

A. INTRODUCTION

For several years we have been studying parameters of serotonin synthesis in rat brain, attempting to discover the roles various components of that process play in the regulation of serotonergic transmission (Mandell, 1974, 1975; Mandell et al., 1974). One such component, the high affinity uptake of tryptophan into serotonergic nerve endings, has proved sensitive to pharmacological manipulation (Knapp and Mandell, 1973). Another, the degree of activity of the rate-limiting enzyme tryptophan hydroxylase in the nerve ending, also changes in response to pharmacological challenge (Knapp et al., 1975), as does the "amount" of tryptophan hydroxylase in the raphé cell bodies and their respective terminal projections (Knapp et al., 1974). Moreover, different drugs cause variations in the latency between the appearance of changes in tryptophan hydroxylase activity in those cell bodies and their appearance in the nerve endings (Knapp and Mandell, 1972a). Finally, some of the mechanisms interact in ways that we venture to call regulatory (Knapp

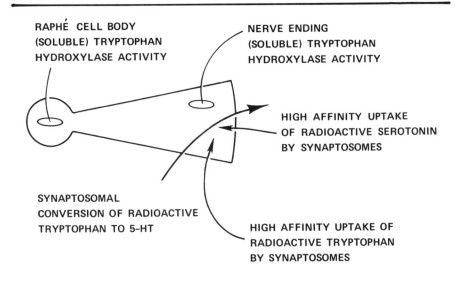

RAPHÉ CELL BODY
(SOLUBLE) TRYPTOPHAN
HYDROXYLASE ACTIVITY

NERVE ENDING
(SOLUBLE) TRYPTOPHAN
HYDROXYLASE ACTIVITY

HIGH AFFINITY UPTAKE
OF RADIOACTIVE SEROTONIN
BY SYNAPTOSOMES

SYNAPTOSOMAL
CONVERSION OF RADIOACTIVE
TRYPTOPHAN TO 5-HT

HIGH AFFINITY UPTAKE OF
RADIOACTIVE TRYPTOPHAN
BY SYNAPTOSOMES

1   Five measures used to assess the capacity of the serotonergic system to synthesize transmitter. Soluble tryptophan hydroxylase assays require the presence of the pteridine cofactor 6-MPH$_4$, dithiothreitol, and exogenous aromatic L-amino acid decarboxylase (Ichiyama et al., 1968, 1970), while the determination of synaptosomal conversion of tryptophan to serotonin does not (Knapp et al., 1974).

and Mandell, 1975). The sketch in Fig. 1 shows the measurements we make when we subject the serotonergic system to a variety of challenges. We have found in general that these processes tend to speed up or slow down differentially, apparently to compensate for the specific effects of the challenges (Mandell et al., 1972, 1974; Mandell, 1975).

In 1972 we began to examine the effects of both short- and long-term administration of lithium on the serotonergic mechanisms described above. We reported that lithium initially stimulated the high affinity transport of radioactive tryptophan into synaptosomes from rat striate cortex, increasing the rate at which the labeled substrate was converted to serotonin in the nerve endings. With continued administration of lithium, soluble tryptophan hydroxylase activity in the midbrain decreased, and later the synaptosomal conversion rate returned to control levels, while the high affinity uptake of substrate remained accelerated (Knapp and Mandell, 1973). We supposed that some feedback from serotonergic receptors, suddenly stimulated by the initial increase in conversion (and release) of transmitter, initiated a decrease in the synthesizing enzyme in the cell bodies of the midbrain raphé, and that the decreased enzyme contingent moved by axoplasmic flow to the nerve endings in the striate cortex. In those early experiments we used the dimethyl pteridine cofactor analogue, 2-amino-4-hydroxy-6,7-dimethyl-5,6,7,8-tetrahydropteridine, but with it we could not measure solubilized enzyme activity in the synaptosomes themselves. Then, with the mono-methyl analogue (2-amino-4-hydroxy-6-methyl-5,6,7,8,-tetrahydropteridine; 6-MPH$_4$) we could determine the effects of both short- and long-term drug administration on solubilized synaptosomal enzyme from the striate cortex. Our results confirmed that 3 weeks of lithium chloride teatment decreased intrasynaptosomal tryptophan hydroxylase activity and kept it reduced; the stimulation of tryptophan uptake into synapto-somes persisted; and our overall (conversion) measure returned to control levels (Knapp and Mandell, 1975).

We have continued working on the acute and chronic effects of lithium on the serotonin biosynthesizing apparatus in the brains of rats. The data we have now suggest that lithium might work by fixing two components of serotonin synthesis at their (opposite) maximal adaptive capacities, thereby allowing transmission to function within a "normal" range. Such buffering of adaptive mechanisms would, theo-retically, preclude further wide variations in transmitter synthesis (and release). In people suffering from bipolar manic-depressive disease, lithium has been found to be equally prophylactic against mania or depression after many months of treatment (Angst et al., 1970; Coppen et al., 1971). Mogens Schou (1957) has made the excellent point that, because of this symmetrical prophylaxis, any viable theory that lithium acts on one neurotransmitter system must explain how the drug can pre-vent extreme fluctuations in transmission in either direction. This caveat would apply to investigations of the stabilization of membrane reactiv-ity (Skou, 1974; Whittam and Ager, 1964), inhibition of catecholamine release (Katz et al., 1968); or alterations in serotonin synthesis (Corrodi et al., 1969; Sheard and Aghajanian, 1970; Ho et al., 1970). Investigating the last possibility, that is, alterations in serotonin synthesis, we have shown that lithium affects the macromolecular mechanisms in that sys-tem in ways that could account for the symmetry of its effectiveness against both extremes of affect. We have also studied the effects of lithium together with the effects of two other drugs that influence specific mechanisms in the same system by treating rats with lithium and then with amphetamine or cocaine, and the results of those experi-ments agree with our model of the way in which lithium might buffer the serotonergic system against extreme increases or decreases in neuro-transmission.

B. METHODS

At specified times before sacrifice, adult male Sprague-Dawley rats (150 to 200 g) were given subcutaneous injections of various doses of lithium chloride, d-amphetamine sulfate, and/or cocaine hydrochloride. Control animals received equimolar sodium chloride in each instance. Although daily doses of 5 to 10 mEq of lithium chloride per kilogram of body weight are reportedly toxic to rats weighing 300 to 400 g (Schildkraut, 1973), within our experimental limits the smaller rats behaved normally with those doses and showed no neurological abnormalities.

Immediately after decapitation brain regions were dissected as follows. The striate cortex (wet weight, 80 mg) included most of the caudate and putamen and some globus pallidus. Midbrain samples (130 mg) extended rostrally up to the mammillary bodies. Medial midbrain samples (37 mg) were obtained by centering the midline in a 2-mm space between cutting wires, leaving the remainder of the midbrain as the lateral midbrain sample (93.5 mg). The medial midbrain sample contained the dorsal and median nuclei of the median raphé (the B7 and B8 nuclear groups of Dahlstrom and Fuxe, 1964); the lateral midbrain, the B9 cell bodies.

Individual regions were homogenized in either 0.32 M sucrose or 1.0 mM Tris buffer, which respectively result in isotonic and hypotonic preparations. When tissue from a single animal was scant, like regions were pooled from animals that had received identical treatment. After homogenates in 0.32 M sucrose were centrifuged at 1000 g the pellets (nuclei and cell debris) were discarded, and the supernates were centrifuged at 12,000 g for 20 min (Gray and Whittaker, 1962). Those supernates were transferred to clean tubes, and the pellets were resuspended in the original volume of homogenization medium. Without delay tryptophan hydroxylase activity was measured in the supernates, and synaptosomal conversion of tryptophan to serotonin was measured in the pellets. Homogenates in 1.0 mM Tris buffer (pH 7.4) were centrifuged at 35,000 g for 20 min. Practically all the measurable enzyme activity from midbrain and striate tissue appeared in the supernates. The hypotonic soluble preparation is distinguished from the solubilized synaptosomal and the isotonic soluble preparations.

Our conversion measure requires maintenance of istonicity to preserve the integrity of the synaptosomes. To measure the enzyme activity component of the conversion measure we resuspended the pellet from 12,000 g centrifugation in 1.0 mM Tris buffer (pH 7.4) and centrifuged at 35,000 g for 30 min. The resulting pellet contained negligible enzymatic activity, and we measured the solubilized synaptosomal enzyme in the supernate.

To measure the pharmacological effects of the drugs under study on tryptophan hydroxylase activity, we assayed the soluble fraction from midbrain samples and the solubilized synaptosomal fraction from striate cortex because soluble enzyme activity has been correlated with serotonergic cell body regions and synaptosomal conversion activity with serotonergic nerve ending regions (Knapp and Mandell, 1972a; Knapp et al., 1974). We also measured synaptosomal conversion of tryptophan to serotonin in striate tissue homogenized in 0.32 M sucrose and fractionated as described above.

The assay for tryptophan hydroxylase (Ichiyama et al., 1968, 1970) couples tryptophan hydroxylase with aromatic amino acid decarboxylase and is possible because the decarboxylase manifests different affinities for tryptophan ($K_m = 1$ mM) and the intermediate product 5-hydroxytryptophan ($K_m = 20$ μM). L-[1-^{14}C]-tryptophan is dissolved in 0.1 M Tris buffer (pH 7.4), and impurities are removed by lyophilization. Our optimal incubation mixture contains 40 μmole Tris-acetate buffer (pH 7.4), 400 nmole 6-MPH$_4$, 700 nmole dithiothreitol, 5 to 10 units decarboxylase (prepared from rat kidney according to the method of Christenson et al., 1970), 100 to 200 μl enzyme preparation (0.3 to 0.6 mg protein), and 4 to 6 nmole labeled tryptophan (15 mCi/mmole) in a final volume of 700 μl. Our substrate concentration is invariably 10 μM (Knapp and Mandell, 1972b).

In the assay for conversion of tryptophan to serotonin the optimal incubation mixture contains 40 μmole Tris-acetate buffer (pH 8.1), 100 to 200 μl enzyme preparation (0.1 to 0.2 mg protein), and 4 to 6 nmole labeled tryptophan (15 mCi/mmole) in a final volume of 700 μl. The necessary cofactors are presumed to be present in the intact synaptosomes.

For either assay, blanks contain boiled enzyme preparation or the appropriate buffer. Before incubation the mixtures are sealed in 15-ml tubes with rubber caps from which are suspended plastic wells containing 100 μl of NCS for the collection of radioactive carbon dioxide. All samples are incubated with shaking for 45 min at 37°C, and the reactions are stopped by the injection of 500 μl of 2 N perchloric acid through the caps. The carbon dioxide accumulates in the NCS over 3 more hours of incubation in the shaker. The wells are then placed directly into counting vials containing 10 ml of a mixture of toluene phosphor (80 ml toluene containing 3.4 g p-bis[2-(5-phenyloxazole)] (PPO) and 0.41 g p-bis[2-(5-phenyloxazolyl)] benzene (POPOP) and absolute ethanol in a 4:1 ratio. Radioactivity is counted in a Beckman LS-250 liquid scintillation spectrophotometer with external standard quench correction. These assays for both tryptophan hydroxylase activity and synaptosomal conversion are linear for as long as 45 min and linear with protein concentration over the range examined. Products of the conversion assay are analyzed by thin-layer chromatography (Knapp et al., 1974).

Aliquots of the intact synaptosomal preparation are used to study the effects of the drugs on substrate uptake. The incubation medium is Krebs-Ringer phosphate buffer containing 100 mM NaCl, 5 mM KCl, 2 mM KH$_2$PO$_4$, 2 mM MgSO$_4$·7H$_2$0, 3.1 mM NaHCO$_3$, 1.5 mM NaPO$_4$ buffer (0.1 M, pH 7), 5.5 mM sodium fumarate, 5 mM sodium pyruvate, 5 mM sodium glutamate, and 12 mM D-glucose (Dawson et al., 1969). In total volumes of 600 to 700 μl there is 0.2 to 0.3 mg protein, and substrate concentrations are varied from 10 to 53 μM, within the range of the high affinity uptake system for tryptophan (Knapp and Mandell, 1972b). Uptake is stopped by dilution with 0.32 M sucrose, and the synaptosomes are collected on Millipore filters (25 mm diameter, 0.65 μ pores) with a Millipore multiple-sampling manifold. Each filter is washed immediately with 2 ml of 0.32 M sucrose, and then radioactivity is counted in 10 ml of toluene phosphor/ethanol in a Beckman LS-200 liquid scintillation counting system. The uptake of L-[3-^{14}C]-tryptophan (50 mCi/mmole) is linear for 5 min at 37°C, with protein concentrations from 0.06 to 0.4 mg.

For comparison we have measured the effects of the drugs on the uptake of radioactive tyrosine or radioactive serotonin in tissues from the same sources. The incubation constituents are the same except for substrate: L-[^{3}H]-tyrosine varies in concentration from 8 to 42 μM; [^{3}H]-serotonin concentration, 0.08 μM.

C. RESULTS

1. Lithium

Lithium chloride (5 to 10 mEq/kg/day) stimulates the uptake of L-[1-^{14}C]-tryptophan by synaptosomes from striate cortex, and the latency of the effect depends on the dose of lithium. At 10 mEq/kg, in 3 days the V_{max} of the uptake reaction increases (Fig. 2). At 5 mEq/kg no changes occur by the third day, but the V_{max} is increased after 21 days of treatment. The K_m values of the uptake hardly change from control values. Our index of uptake is the radioactivity retained after the synaptosomes are incubated with radioactive substrate. Enhancement of V_{max} would be consonant with an increase in the relative velocity of the uptake of radioactive substrate. Lithium has been reported to inhibit catecholamine release (Katz et al., 1968) so we preincubated synaptosomes from rats treated with lithium chloride or sodium chloride for 5 min with L[3-^{14}C]-tryptophan (10 μM) and found that the efflux of radioactivity over 10 min from the two preparations was comparable (Knapp and Mandell, 1973). Thus we ruled out the possibility that our data reflected inhibition of serotonin release or facilitated exchange of endogenous for exogenous tryptophan.

In vitro, with high concentrations of lithium not intended to reflect mechanisms *in vivo* (where time and dose seem to interact), we compared the effects of lithium on the uptakes of radioactive tyrosine and tryptophan. The effect on tryptophan uptake appears to be relatively specific, and, moreover, dependent on the level of lithium present. At a concentration of 10 mM, lithium chloride stimulates the high affinity uptake of labeled tryptophan; at a concentration of 50mM it stimulates the uptake further. The K_m value of the uptake does not change, but the V_{max} rises 45% with the higher concentration of lithium. Under control conditions at equimolar concentrations of substrate, tyrosine uptake rises only 20% in the presence of the same concentration of lithium.

The effects of lithium chloride on striate synaptosomal conversion of radioactive tryptophan to serotonin are functions of both dose and time (Fig. 3). High doses of lithium (10 mEq/kg) result in 30 to 40% increases in conversion as well as uptake (Figs. 2 and 3). At sacrifice the rats manifested serum lithium levels of 0.6 ± 0.1 mEq/liter (Zettner et al., 1968). *In vitro* comparable increments in either uptake or conversion demand much higher levels of the drug. Despite a large disparity between the specific activities of conversion and high affinity uptake (5 to 7 versus 350 to 450 pmoles/mg/min) with substrate at 10 μM and identical doses of lithium chloride, we have demonstrated changes of similar proportion between the two parameters with other manipulations (Knapp and Mandell, 1972b, 1973; Knapp et al., 1974). Different absolute values, but proportional fluctuations in the two processes might occur if there were a preferential pathway for substrate transport and synthesis, or a high affinity uptake system in cells other than serotonergic neurons (Kuhar et al., 1972; Mandell et al., 1974), or if the efflux of unconverted tryptophan resulted in relatively low net hydroxylation of the substrate.

Soluble enzyme activity in the isotonic preparation from midbrain begins to decrease on the first day of treatment with lithium chloride (10 mEq/kg; Fig. 4). On the third day we observe a maximum decrement of around 30%, which remains on the fifth and seventh days. With 5 mEq/kg of body weight, 10 days af administration pass before a significant decrement appears. After 21 days of treatment, that decrement has reached 30%. Interactions between dose and time are implicit in the effects of lithium treatment on midbrain soluble tryptophan hydroxy-

THE EFFECT OF
SHORT-TERM LiCl TREATMENT
(10 meq kg⁻¹ 3 days⁻¹) ON ¹⁴C - TP
UPTAKE INTO RAT STRIATE
SYNAPTOSOMES

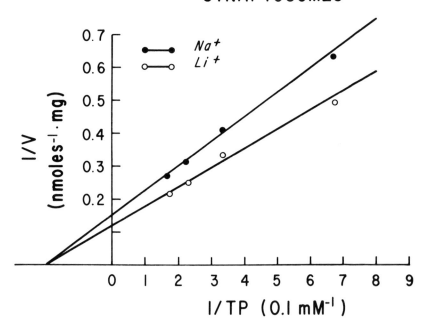

2 The effects of lithium chloride (10 mEq/kg/day × 3) on the uptake of L-[3-¹⁴C]-tryptophan into rat striate synaptosomes. Velocity is plotted as nmole/labeled substrate retained/mg protein/4 min. Substrate concentrations, 15 to 53 μM. With Na⁺,. V_{max} was 6.8 ± 5 and K_m was 50 ± 1.5 μM; with Li⁺, V_{max} was 9.0 ± 1 and K_m was 50 ± 1 μM, estimated according to Wilkinson (1961). (Knapp and Mandell, 1975).

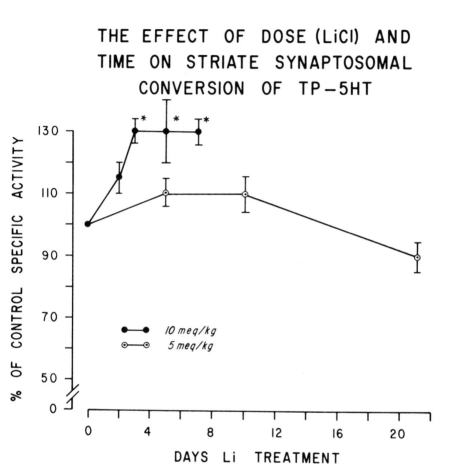

THE EFFECT OF DOSE (LiCl) AND TIME ON STRIATE SYNAPTOSOMAL CONVERSION OF TP−5HT

10 meq/kg
5 meq/kg

% OF CONTROL SPECIFIC ACTIVITY

DAYS Li TREATMENT

3

Time course of the effects of two doses of lithium chloride on striate synaptosomal conversion of tryptophan to serotonin. Controls received equivalent doses of NaCl. Animals were sacrificed 24 hr after the final daily subcutaneous drug administration. Control activity = 200 ± 10 pmole/mg/45 min (*$p < 0.01$) (Knapp and Mandell, 1975).

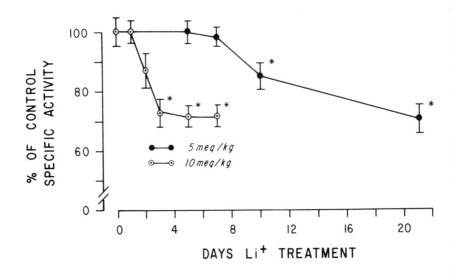

THE EFFECT OF DOSE (LiCl) AND TIME ON MB SOLUBLE TP-OHase

4 Time course of the effects of two doses of lithium chloride on midbrain soluble tryptophan hydroxylase activity. Controls received equivalent doses of NaCl. Animals were sacrificed 24 hr after the final daily subcutanous drug administration. Control activity = 500 ± 15 pmole/mg/45 min (*$p < 0.01$) (Knapp and Mandell, 1975).

lase activity, as they are in the uptake and conversion activities. *In vitro*, concentrations of the drug ranging from 1 to 100 mM have no effect on the activity of the isotonic preparation of soluble enzyme from midbrain.

The kinetics of the reduced enzyme activity with respect to the pteridine cofactor show a decrease in V_{max} rather than a change in K_m value, which suggests that the decrement could come from a reduction in the number of available enzyme sites or molecules. Although the decrement appears to move by axonal flow from the raphé nuclei to the striate cortex, it is not clear whether lithium decreases the synthesis or increases the turnover of tryptophan hydroxylase in the cell bodies. We do suspect that the effect is related to the response of the serotonergic receptors to the initial increments in transmitter synthesis.

The simultaneous effects of short-term (3 days) administration of lithium chloride (10 mEq/kg) on four serotonergic measures are shown in Fig. 5. Both uptake and conversion of labeled tryptophan to serotonin begin their rise within the first day, but soluble enzyme activity from neither cell bodies in the midbrain nor striate synaptosomes in hypotonic preparation shows a change the first day. On the second day increases in uptake and conversion are not apparent, but the enzyme activity from the cell bodies and nerve endings begins to decrease. On the third day the uptake of substrate and its conversion to transmitter increase further. After 21 daily treatments with lithium chloride (5 mEq/kg; Fig. 6) the decrement in cell body and nerve ending enzyme activity is still present, as is the increase in the uptake of labeled substrate, but conversion of labeled substrate to transmitter has returned to control levels.

Perez-Cruet, Tagliamonte, Tagliamonte, and Gessa (1971) have shown that treating rats with lithium carbonate increases the level of tryptophan in their sera and brains. We studied the effects on conversion activity of progressively larger doses of L-tryptophan, and found that conversion of labeled substrate increased as a function of dose. The addition of unlabeled tryptophan would, if anything, dilute the radioactivity of the conversion product, so we may have underestimated the actual increase in conversion. Then, to find out whether the increase in conversion had been preceded by an increase in uptake of radioactive substrate caused by the exogenous tryptophan, we preincubated striate synaptosomes with 100 μM L-tryptophan and examined the uptake kinetics. The V_{max} increased, but the K_m value did not, which suggested that the exogenous tryptophan increased the uptake of radioactive substrate by increasing the number of uptake sites. We do not know whether this tryptophan-induced increase in uptake and conversion accounts for the initial effects of lithium. Obviously the concentration of tryptophan *in vivo* would be critical, and we do not know whether tryptophan loading can bring substrate to the appropriate levels at the relevant sites. Others' attempts to determine regional and subcellular levels of tryptophan after administering exogenous tryptophan are not conclusive (Perez-Cruet et al., 1971; Fernstrom and Wurtman, 1971; Grahame-Smith, 1971).

2. *D*-Amphetamine

d-Amphetamine almost immediately reduces synaptosomal conversion of tryptophan to serotonin in striate nerve endings, and that reduction is associated with a decrease in intrasynaptosomal and lateral mid-

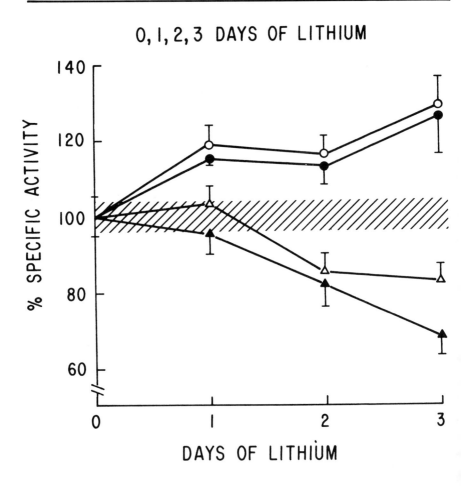

0, 1, 2, 3 DAYS OF LITHIUM

% SPECIFIC ACTIVITY

DAYS OF LITHIUM

5   Simultaneous effects of three daily lithium chloride injections (10 mEq/kg) on °——° high affinity uptake of labeled tryptophan (control = 1200 ± 60 pmole/mg/4 min), •——• striate synaptosomal conversion (control = 250 ± 12 pmole/mg/45 min); △——△ midbrain soluble enzyme activity (control = 525 ± 20 pmole/mg/45 min); ▲——▲ striate solubilized enzyme activity (control = 150 ± 5 pmole/mg/45 min). Substrate concentrations were 10 μM for uptake, 8 μM for the assays for all measures after 3 days of lithium (p < 0.05) (Knapp and Mandell, 1975).

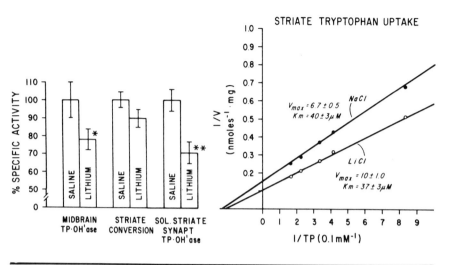

6 Simultaneous effects of 21 daily lithium chloride injections (5 mEq/kg) on midbrain soluble enzyme (control = 475 ± 20 pmole/mg/45 min, $p < 0.05$); striate synaptosomal conversion (control = 200 ± 10 pmole/mg/45 min); striate solubilized enzyme (control = 160 ± 5 pmole/mg/45 min. $p < 0.001$); high affinity uptake, plotted as nmole labeled substrate retained/mg/4 min. Substrate concentration for uptake ranged from 12 to 53 μM. V_{max} and K_m are estimated according to Wilkinson (1961). (Knapp and Mandell, 1975.)

brain (distinguished from medial midbrain) tryptophan hydroxylase activity. Within an hour the specific activities of all three measures are reduced about 40% (Knapp et al., 1974; Fig. 7).

We administered an amphetamine challenge to rats that had been treated with lithium daily for 3 days (at 10 mEq/kg) or for 6, 9, or 16 days (at 5 mEq/kg), and compared two parameters of serotonin synthesis with those in matched controls. The amphetamine challenge was administered at 5 mg/kg 22 hr after the last lithium injection and 2 hr before sacrifice. With 3 days of lithium pretreatment the amphetamine had no effect on (did not decrease) the lithium-induced increase in substrate conversion. With 9 days of lithium pretreatment amphetamine still failed to decrease the lithium-augmented conversion and also failed to reduce tryptophan hydroxylase activity. With 16 days of lithium pretreatment, there was significantly less percent change in either parameter than there had been after 3 days (Table I).

Table I. Decreasing Responses to Amphetamine by
Serotonin-Synthesizing Mechanisms with
Long-Term Lithium Treatment [a]

| | Percent change in amphetamine effect [b] | |
Days of lithium treatment	Synaptosomal conversion TRP → serotonin	Raphé-soluble TRP hydroxylase activity
3	− 32	+ 22
6	− 18	+ 10
9	− 11	− 3
16	− 9	+ 4

[a] NaCl or LiCl was administered subcutaneously for 3 days
(10 mEq/kg), 6, 9, or 16 days (5 mEq/kg). N = 8 in each
group. Rats were sacrificed 24 hr after the last injection
of NaCl or LiCl and 2 hr after an injection of d-ampheta-
mine sulfate (5 mg/kg). Control conversion specific
activity (NaCl · amphetamine = 100%) was 200 ± 11
pmole/mg protein/45 min; control raphé-soluble enzyme
activity was 750 ± 50 pmole/mg/45 min.

[b] $\dfrac{\text{(NaCl} \times \text{amphetamine)} - \text{(LiCl} \times \text{amphetamine)}}{\text{(NaCl} \times \text{amphetamine)}} \times 100$

3. Cocaine

We reported in 1972 (Knapp and Mandell, 1972b) that *in vitro*
cocaine inhibited the high affinity uptake of tryptophan ($K_m = 10$ μM)
as drug concentrations increased. On the other hand, although it reduced
soluble tryptophan hydroxylase activity, morphine did not affect synaptosomal uptake of substrate. Thus by different means both narcotic drugs
reduced the conversion of labeled tryptophan to serotonin by striate
synaptosomes. Also, tryptophan hydroxylase activity from cell body
regions was increased after cocaine (Fig. 8).

In our subsequent experiments, a large dose of cocaine hydrochloride
in vivo (100 mg/kg) reduced striate synaptosomal conversion of tryptophan to serotonin within an hour, and we noted a compensatory increase
in the activity of tryptophan hydroxylase solubilized from striate nerve
endings. There was no effect on serotonin uptake (Fig. 9). The increase
in enzyme activity appeared to originate in the cell bodies of the
lateral midbrain, peaked after 1 hour, and had subsided an hour later
(Fig. 10). Figure 11 further illustrates the anatomic specificity within
the midbrain raphé. Enzyme activity was determined in both the medial
(B7 and B8) and the lateral (B9) nuclei with two homogenization

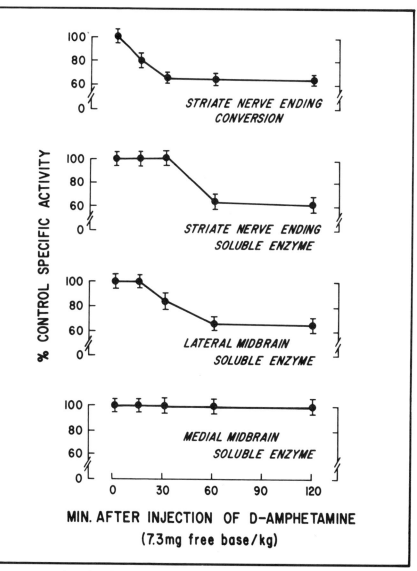

7

Acutely after the injection of d-amphetamine, tryptophan-to-serotonin conversion activity, lateral midbrain tryptophan hydroxylase activity, and striate synaptosomal solubilized enzyme activity were significantly reduced below their respective control levels. Control velocities in pmole/mg protein/45 min; striate conversion = 220 ± 10; striate synaptosomal enzyme 150 ± 7; lateral midbrain soluble enzyme = 206 ± 8; medial midbrain soluble enzyme = 98 ± 40.

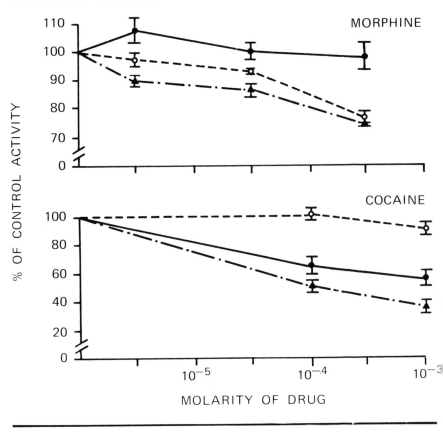

8 *In vitro* effects of morphine and cocaine on ——•—— uptake of tryptophan by septal synaptosomes (control = 450-500 pmole/mg/5 min); - - - ○ - - - midbrain soluble tryptophan hydroxylase (control = 100 pmole/mg/45 min); •—▲•— synaptosomal conversion of tryptophan to serotonin (control = 125 pmole/mg/45 min). Morphine reduced enzyme and conversion activities, and cocaine reduced substrate uptake and conversion ($p < 0.05$) (Knapp and Mandell, 1972b).

media, one that leaves the cells intact (0.32 M sucrose) and one that lyses them (1.0 mM Tris buffer). The increase in tryptophan hydroxylase after cocaine was seen only in the area encompassing the B9 serotonergic cell bodies.

In Fig. 12 we have graphed the effects of cocaine, of lithium, and of the two drugs together, administered as described above. Whereas cocaine alone reduced high affinity uptake of radioactive tryptophan and lithium alone stimulated the same mechanism, after 3 days of lithium treatment (10 mEq/kg), cocaine (100 mg/kg) did not significantly

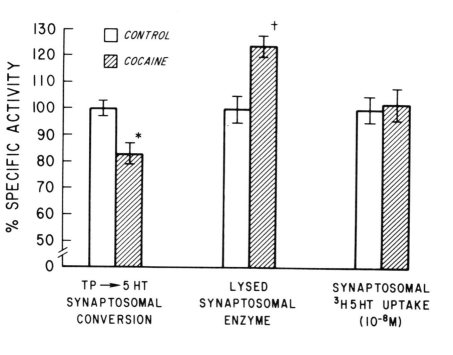

9

Effects of cocaine hydrochloride (100 mg/kg) administered 1 hr before sacrifice on striate synaptosomal trytophan-to-serotonin conversion, intrasynaptosomal tryptophan hydroxylase activity, and synaptosomal uptake of labeled serotonin (5-HT), graphed as percentages of their respective control specific activities: 225 ± 11.5, 75 ± 3.5 pmole/mg/30 min, and 1.19 ± 0.05 pmole/mg/4 min. Substrate concentration was 10 μM for enzyme assays; 0.08 μM for ^{3}H-5-HT uptake determination. Cocaine significantly reduced conversion activity and significantly increase soluble enzyme activity ($^*p < 0.001$; $^+p < 0.032$).

affect uptake. The results were parallel with regard to the conversion of radioactive substrate to transmitter in striate synaptosomes: Cocaine decreased conversion; lithium increased it; and after 3 days of lithium treatment, cocaine did not significantly affect it. Cocaine induced increases in tryptophan hydroxylase activity in both lateral midbrain and striate cortex; lithium decreased enzyme activity in the midbrain and in striate cortex; after 3 days of lithium treatment, cocaine did not significantly affect enzyme activity in either place.

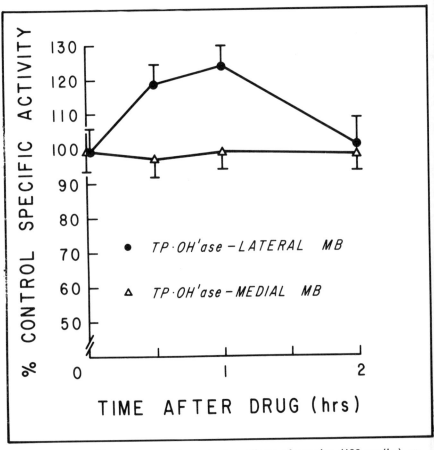

10 Time course of the selective effects of cocaine (100 mg/kg) on ——●—— lateral midbrain versus ——△—— medial midbrain serotonergic cell bodies (B9 versus B7 and B8). There was no detectable cocaine-induced change in tryptophan hydroxylase prepared from medial midbrain (control = 750 ± 25.8 pmole/mg/30 min, but enzyme from lateral midbrain was increased 30 and 60 min after drug treatment (control = 125 ± 5 pmole/mg/30 min, $p < 0.005$).

D. DISCUSSION

Latency to effect is a major puzzle with respect to psychotherapeutic drug action. Treating mania with lithium may involve a delay of 5 to 10 days before the clinical effects are substantial, and prophylaxis against both mania and depression takes many months to develop (Cade, 1949; Schou, 1957; Freyhan, 1971). Perhaps initial doses do not promptly reach effective levels in the brain or blood. Schou (1973a,b) and others have reported that brain levels of lithium achieve a steady state in ani-

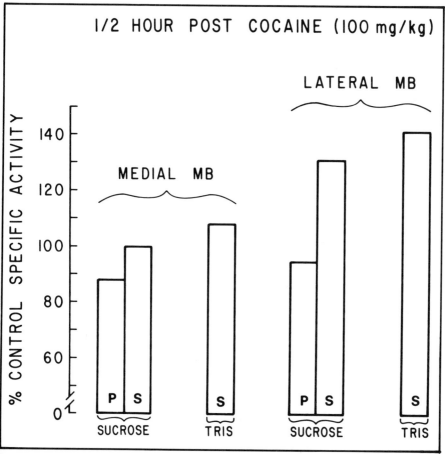

Cocaine-induced effects ½ hr after injection on medial and lateral midbrain (MB) tryptophan hydroxylase prepared by isotonic homogenization in 0.32 M.. sucrose and centrifugation at 12,000 g (which keeps synaptosomes intact) and by hypotonic homogenization in 1.0 mM Tris buffer (pH 7.4) and centrifugation at 35,000 g (which destroys synaptosomal integrity). Data are presented as mean ($N = 6$) percentages of corresponding control activities from saline-treated animals. Of the sucrose preparations from medial MB, control activity in pellet (P) = 139 ± 5 pmole/mg/30 min; in supernate (S), 750 ± 25.2. Of the sucrose preparations from lateral MB, control activity in P was 51.5 ± 3.0 pmole/mg/30 min; in S 90.6 ± 4.0. Of the Tris preparations, control activity was 770 ± 20 pmole/mg/30 min in medial and 103.6 ± 4 pmole/mg/30 min in lateral MB. There was a mean decrease in synaptosomal conversion in the pellets from both midbrain areas: The activity in neither supernate from medial MB changed significantly after cocaine; both soluble enzyme measures from lateral MB rose significantly after cocaine ($p < 0.005$ for sucrose preparation; $p < 0.013$ for Tris preparation).

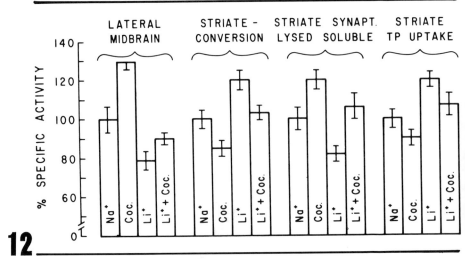

12

The independent effects of cocaine (100 mg/kg) and lithium (10 mEq/kg) and the effects of cocaine after 3 days of lithium pretreatment on four parameters of serotonin synthesis in rat brain. Na+ represents values from animals pretreated with NaCl. Data are given as percentages of the following control specific activities: lateral midbrain soluble tryptophan hydroxylase activity = 200 ± 5 pmole/mg/30 min; striate conversion = 180 ± 7 pmole/mg/30 min; striate synaptosomal lysed soluble enzyme = 111 ± 4.5 pmole/mg/30 min; striate uptake of L-[3-^{14}C]-tryptophan, 1150 ± 60 pmole/mg/5 min. Independently, cocaine and lithium altered lateral midbrain and striate enzyme activities ($p < 0.028$) as well as striate conversion of tryptophan to serotonin ($p < 0.05$). Lithium increased radioactive tryptophan uptake, whereas cocaine decreased it. Pretreatment with lithium chloride antagonized the cocaine-induced changes in serotonin biosynthesis reflected by all four measures.

mals only after five to seven daily drug administrations. On the other hand, in maniac patients large initial doses of lithium speed the onset of its effect hardly at all, but magnify its side effects greatly (Schou, personal communication). It is difficult to believe that plasma or brain blood levels of drug account for all the latency to lithium's effectiveness. Is it not possible then that some aspects of clinical efficacy are due to secondary or tertiary mechanisms activated by the primary drug effect? Conventional tolerance usually does not develop with psychotherapeutic drugs like the phenothiazines, tricyclic antidepressants, or lithium, although with chronic administration many side effects diminish or vanish. The therapeutic effects of lithium do not usually require progressively larger doses, in fact the same dose can retain, if not increase, its clinical efficacy over years. Must we not consider the possibility that treatment with lithium could actually be treatment by the induction of tolerance mechanisms?

No reliable theory correlates increases or decreases in serotonin synthesis *per se* to either mania or depression. Serotonergic transmission may, rather, serve a regulatory, not a direct, role in relation to emotions and behavior. Thus it may be important that the sequence of events we report with the chronic administration of lithium affects two adaptive mechanisms—one that can increase serotonin synthesis and one that can decrease it. The control levels of nerve ending conversion of tryptophan to serotonin with chronic lithium treatment are, we believe, the net effect of continued enhancement of substrate uptake and continued reduction in tryptophan hydroxylase activity (Fig. 13). By other manipulations we can induce each change independently. For example, amphetamine, cocaine, and tryptophan loads, respectively, affect various constituents of serotonin synthesis, but none of these changes occurs when the challenge follows lithium treatment. It is as though lithium fixes the uptake of tryptophan at its maximum and the intraneuronal rate-limiting enzyme activity at its minimum, and exhausts the overall adaptive capacity, having returned the synthesis of transmitter to "normal" and left the system buffered against further extremes in fluctuation.

Goodwin, Sack, and Post (1975) have reported clinical data that suit our hypothesis nicely. Using probenecid blockade of metabolite egress from the cerebrospinal fluid, they demonstrated a biphasic response in the turnover of the serotonergic metabolite 5-hydroxyindoleacetic acid in the spinal fluid of depressed patients being treated with lithium carbonate. After 5 days of lithium treatment, 5-hydroxyindoleacetic acid was elevated, which could correspond to the increase we note in tryptophan uptake and conversion to serotonin; after 21 days of treatment levels of the metabolite were again at or below control levels.

Our model implies too that regular tryptophan loads might be effective in bipolar affective disorder. Increased circulating tryptophan might increase uptake of substrate into the nerve endings by mass action (whereas lithium stimulates the uptake mechanism), leading to increased transmitter synthesis and a reflexive decrease in tryptophan hydroxylase, which would return serotonin synthesis to "normal," narrowing the range of potential adaptive changes. Nathan Kline (personal communication) has successfully substituted daily tryptophan loads for lithium in patients with affective disease who showed marked intolerance to lithium, and Murphy, Baker, Goodwin, Miller, Kotin, and Bunney (1974) have reported both antidepressant and antimanic effects when they gave tryptophan to patients with bipolar affective disorder.

In view of its biphasic effects on the serotonergic system, perhaps lithium's apparent antagonism of the effects of amphetamine could be

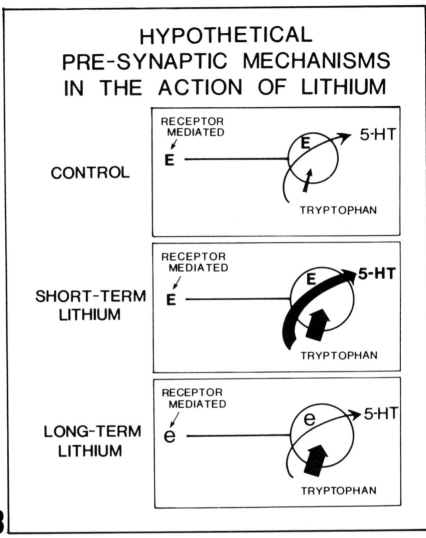

HYPOTHETICAL PRE-SYNAPTIC MECHANISMS IN THE ACTION OF LITHIUM

CONTROL

RECEPTOR MEDIATED

E → 5-HT

TRYPTOPHAN

SHORT-TERM LITHIUM

RECEPTOR MEDIATED

E → 5-HT

TRYPTOPHAN

LONG-TERM LITHIUM

RECEPTOR MEDIATED

e → 5-HT

TRYPTOPHAN

13

Hypothetical presynaptic mechanisms in the action of lithium. In the control neuron, tryptophan hydroxylase (E) is optimal in both cell body and nerve ending. Tryptophan is taken up through the neuronal membrane and converted by E to serotonin, which is released from the nerve ending. After short-term lithium treatment (cf. Fig. 2) tryptophan uptake is augmented and consequently synthesis and release of serotonin are increased, since intraneuronal enzyme is not saturated with regard to substrate. After long-term lithium treatment, "amount" of enzyme has been reduced (E → e) to compensate for the enhanced bombardment of the receptor. Tryptophan uptake is still augmented, but serotonin synthesis and release at the nerve ending have returned to control levels because of the enzyme deficit (e instead of E).

explained as follows. If serotonin is a nonspecific inhibitor of motor activity, as is generally agreed (Breese et al., 1974; Mabry and Campbell, 1973), both spontaneous motor activity and amphetamine-induced hyperactivity would be suppressed by an increase in functional serotonin such as lithium induces by increasing substrate and conversion (and release) initially. Segal, Callaghan, and Mandell (1975) showed that amphetamine-induced hyperactivity in rats is markedly attenuated with 1 day of lithium pretreatment, just as spontaneous motor activity is depressed with lithium alone. With longer pretreatment (8 days) spontaneous motor activity returns to control levels, but amphetamine-induced hyperactivity is still significantly reduced. By our model, amphetamine's usual acute release of serotonin and rapid compensatory reduction of its synthesis might not appear with short-term lithium pretreatment, being counteracted by the early stimulatory effects of lithium on transmitter synthesis. The reduction in tryptophan hydroxylase that longer-term treatment with lithium induces and maintains could prevent amphetamine's release of serotonin from provoking a compensatory decrease in enzyme activity (corresponding in time to the return of both spontaneous motor activity and conversion of tryptophan to serotonin to control levels). That is, as lithium treatment reduces tryptophan hydroxylase there would be less serotonin available for amphetamine to release and therefore less change in synaptosomal conversion of tryptophan to serotonin induced by an amphetamine challenge. Moreover, compensatory decreases in tryptophan hydroxylase after amphetamine would progressively decline. All this of course implies a relationship between cell body enzyme change and receptor sensitivity.

Our studies show that both amphetamine and cocaine act specifically on B9 cell bodies, a finding which Harvey, McMaster, and Yunger (1975) have also reported with p-chloroamphetamine. The specificity cannot be attributed to the drugs acting on short-axoned neurons because the anatomical differentiation appeared with both intact and lysed (isotonic and hypotonic) preparations from both regions. Thus B9 cell bodies appear to be uniquely implicated in the biochemical changes we have measured in the striate cortex, which is consistent with demonstrations of B9 projection to the striate cortex in the rat by histofluorescence techniques (Fuxe and Jonsson, 1967).

Because cocaine's independent effects on these mechanisms are the opposite of lithium's independent effects, their neurobiological antagonism can partially confirm the actions of both drugs. Lithium's demonstrated antagonism of the behavioral effects of amphetamine (Segal et al., 1975; Furukawa, 1975; Flemenbaum, 1974; Cassens and Mills,

1973), morphine (Carroll and Sharp, 1971), and alcohol (Pendery and Huey, 1974), all drugs which, like cocaine, reduce the biosynthesis of serotonin and produce states of excitement in man, may imply similar behavioral correlates for its antagonism of the biochemical effects of cocaine.

Much more basic and clinical research needs to be done before we can know certainly whether lithium works as we speculate here on the basis of our studies showing a kind of "cross-tolerance" for drugs that induce model psychopathological states. Two maniac-depressive patients in treatment with lithium have recently told us that illicitly obtained cocaine failed to affect them when they used it in the presence of others who experienced the expected effects from the drug. It is tempting to wonder whether lithium might be useful in the treatment of compulsive stimulant users.

ACKNOWLEDGMENT

This work is supported by United States Public Health Service Grant DA-00265.

E. REFERENCES

Angst, J., Weis, P., Grof, P., Baastrup, P., and Schou, M. (1970). Lithium prophylaxis in recurrent affective disorder. *Brit. J. Psychiat.* 116, 604–614.

Breese, G. R., Cooper, B. R., and Mueller, R. A. (1974). Evidence for an involvement of 5-hydroxytryptamine in the action of amphetamine. *Brit. J. Pharmacol.* 52, 307–319.

Cade, J. F. J. (1949). Lithium salts in the treatment of psychotic excitement. *Med. J. Australia* 36, 349–352.

Carroll, B. J., and Sharp, B. P. (1971). Rubidium and lithium: Opposite effect on amine-mediated excitement. *Science* 172, 1355–1357.

Cassens, G. P., and Mills, A. W. (1973). Lithium and amphetamine: Opposite effects on threshold of intracranial reinforcement. *Psychopharmacologia* 30, 283–290.

Christenson, J. G., Dairman, W., and Udenfriend, S. (1970). Preparation and properties of a homogeneous aromatic L-amino acid decarboxylase from hog kidney. *Arch. Biochem. Biophys.* 141, 356–367.

Coppen, A. Noguera, R., Bailey, J. *et al.* (1971). Prophylactic lithium in affective disorders. *Lancet* II, 275–279.

Corrodi, H., Fuxe, K., and Schou, M. (1969). The effect of prolonged lithium administration on cerebral monoamine neurons in the rat. *Life Sci.* 8I, 643–651.

Dahlstrom, A., and Fuxe, K. (1964). Evidence for the existence of monoamine-containing neurons in the central nervous system. *Acta Physiol. Scand.* 62: *Suppl.* 232, 1–55.

Dawson, R. M. C., Elliott, D. C., Elliott, W. H., and Jones, K. M., eds. (1969). *Data for Biochemical Research,* Oxford University Press, Cambridge.

Fernstrom, J. D., and Wurtman, R. J. (1971). Brain serotonin content: Physiological dependence on plasma tryptophan levels. *Science* 173, 149–152.

Flemenbaum, A. (1974). Does lithium block the effects of amphetamine? A report of three cases. *Amer. J. Psychiat.* 131, 820–821.

Freyhan, F. A. (1971). Lithium treatment: Prophylactic or compensatory? *Amer. J. Psychiat.* 128, 122.

Furukawa, T., (1975). Modification by lithium of behavioral responses to methamphetamine and tetrabenazine. *Psychopharmacologia* 42, 243–248.

Goodwin, F. K., Sack, R. L., and Post, R. M. (1975). Clinical evidence for neurotransmitter adaptation in response to antidepressant therapy. In: *Neurobiological Mechanisms of Adaptation and Behavior* (A. J. Mandell, ed.), Raven press, New York, pp. 33–45.

Fuxe, K., and Jonsson, G. (1967). A modification of the histochemical fluorescence method for the improved localization of 5-hydroxytryptamine. *Histochemie* 11, 161–166.

Grahame-Smith, D. G. (1971). Studies *in vivo* on the relationship between brain tryptophan, brain 5-HT synthesis, and hyperactivity in rats treated with a monoamine oxidase inhibitor and l-tryptophan. *J. Neurochem.* 18, 1053–1066.

Gray, E. G., and Whittaker, V. P. (1962). The isolation of nerve endings from brain: An electron microscopic study of cell fragments derived by homogenization and centrifugation. *J. Anat.* 96, 71-88.

Harvey, J. A., McMaster, S. E., and Yunger, L. M. (1975). *p*-Chloroamphetamine: Selective neurotoxic action in brain. *Science* 187, 841-843.

Ho, A. K. S., Loh, H. H., Craves, F., Hitzemann, R. J. and Gershon, S. (1970). The effect of prolonged lithium treatment on the synthesis rate and turnover of monoamines in brain regions of rats. *Eur. J. Pharmacol.* 10, 72–78.

Ichiyama, A., Nakamura, S., Nishizuka, Y., and Hayaishi, O. (1968). Tryptophan hydroxylase in mammalian brain. *Adv. Pharmacol.* 6A, 5–17.

Ichiyama, A., Nakamura, S., Nishizuka, Y., and Hayaishi, O. (1970). Enzymic studies on the biosynthesis of serotonin in mammalian brain. *J. Biol. Chem.* 245, 1699–1709.

Katz, R. I., Chase, R. N., and Kopin, I. J. (1968). Evoked release of norepinephrine and serotonin from brain slices: inhibition by lithium. *Science* 162, 466–467.

Knapp, S., and Mandell, A. J. (1972a). Parachlorophenylalanine—its three phase sequence of interactions with the two forms of brain tryptophan hydroxylase. *Life Sci.* 11, 761–771.

Knapp, S., and Mandell, A. J. (1972b). Narcotic drugs: Effects on the serotonin biosynthetic systems of the brain. *Science* 177, 1209–1211.

Knapp, S., and Mandell, A. J. (1973). Short and long-term lithium administration: Effects on the brain's serotonergic biosynthetic systems. *Science* 180, 645–647.

Knapp, S., and Mandell, A. J. (1975). Effects of lithium chloride on parameters of biosynthetic capacity for 5-hydroxytryptamine in rat brain. *J. Pharmacol. Exp. Ther.* 193, 812–823.

Knapp, S., Mandell, A. J., and Bullard, W. P. (1975). Calcium activation of brain tryptophan hydroxylase. *Life Sci.* 16, 1583–1594.

Knapp, S., Mandell, A. J., and Geyer, M. A. (1974) Effects of amphetamine on regional tryptophan hydroxylase activity and synaptosomal conversion of tryptophan to 5-hydroxytryptamine in rat brain. *J. Pharmacol. Exp. Ther.* 189, 676–689.

Kuhar, M. J., Roth, R. H., and Aghajanian, G. K. (1972). Synaptosomes from forebrains of rats with midbrain raphé lesions: Selective reduction of serotonin uptake. *J. Pharmacol. Exp. Ther.* 181, 36–45.

Mabry, P. D., and Campbell, B. A. (1973). Serotonergic inhibition of catecholamine-induced behavioral arousal. *Brain Res.* 49, 381–391.

Mandell, A. J. (1974). The role of adaptive regulation in the pathophysiology of psychiatric disease. *J. Psychiat. Res.* 11, 173–179.

Mandell, A. J. (1975). Neurobiological mechanisms of presynaptic metabolic adaptation and their organization: Implications for a pathophysiology of the affective disorders. In: *Neurobiological Mechanisms of Adaptation and Behavior* (A. J. Mandell, ed.), Raven Press, New York, pp. 1–32.

Mandell, A. J., Segal, D. S., Kuczenski, R. T., and Knapp, S. (1972). Some macromolecular mechanisms in CNS neurotransmitter pharmacology and their psychobiological organization. In: *Chemistry of Mood, Motivation, and Memory* (J. L. McGaugh, ed.), Plenum, New York, pp. 105–148.

Mandell, A. J., Knapp, S., and Hsu, L. L. (1974). Some factors in the regulation of central serotonergic synapses. *Life Sci.* 14, 1–17.

Pendery, M., and Huey, L. (1974). In: *Highlights, 19th Annual Conf., Studies in Mental Health and Behav. Sci.*, US Vet. Adm., New Orleans.

Perez-Cruet, J., Tagliamonte, A., Tagliamonte. P., and Gessa, G. L. (1971). Stimulation of serotonin synthesis by lithium. *J. Pharmacol. Exp. Ther.* 178, 325–330.

Schildkraut, J. J. (1973). The effects of lithium on biogenic amines. In: *Lithium: Its Role in Psychiatric Research and Treatment* (S. Gershon and B. Shopsin, eds.), Plenum, New York, pp. 51–73.

Schou, M. (1957). Biology and pharmacology of the lithium ion. *Pharmacol. Rev.* 9, 17–58.

Schou, M. (1973a). Possible mechanisms of action of lithium salts: Approaches and perspectives. *Biochem. Soc. Trans.* 1, 81–87.

Schou, M. (1973b). Prophylactic lithium maintenance treatment in recurrent endogenous affective disorders. In: *Lithium: Its Role in Psychiatric*

Research and Treatment. (S. Gershon and B. Shopsin, eds.), Plenum, New York.

Segal, D. S., Callaghan, M., and Mandell, A. J. (1975). Alterations in behavior and catecholamine synthesis induced by lithium. *Nature* (London) 254, 58–59.

Sheard, M. H., and Aghajanian, G. K. (1968). Stimulation of midbrain raphe neurons: Behavioral effects of serotonin release. *Life Sci.* 7, 19–25.

Skou, J. C. (1974). Enzymatic aspects of active linked transport of Na^+ and K^+ through the cell membrane. *Prog. Biophys. Mol. Biol.* 14, 133–166.

Whittam, R., and Ager, M. E. (1964). Vectorial aspects of adenosine-triphosphatase activity in erythrocyte membranes. *Biochem. J.* 93, 337–348.

Wilkinson, G. N. (1961). Statistical estimation in enzyme kinetics. *Biochem, J.* 80, 324–332.

Zettner, A., Rafferty, K., and Jarecki. H. (1968). Determination of lithium in serum and urine by atomic absorption spectroscopy. *Atomic Absorption Newslett.* 7, 32–34.

Index

A

Acetylcholine
   choline as precursor, 168
   effect on brain mitochondrial activity,
      158-163
   effect on redox potential, 157-163
Acyl carrier protein
   of Escherichia coli, 217
5'-Adenosine monophosphate (5'-AMP)
   effect of ethanol, 176, 177
   substrate for ecto-5'-nucleotidase,
      176, 177
   substrate for total 5'-nucleotidase,
      176, 177
Adenosine triphosphatase (ATP-ase)
   as membrane-associated enzyme,
      170-174, 178, 179
   conformational change, 171, 174
   effect of acute ethanol treatment,
      170-174, 178, 179
   effect of chronic ethanol treatment,
      170-174, 178, 179
   in brain cultures, 170-174, 178, 179
   in mouse, 170
   in rat, 170

Mg^{++}-ATP-ase, 170-174, 178
Na^+, K^+-ATP-ase, 170-174, 178
Na^+, K^+-ATP-ase as a glycoprotein, 174
   of astroblasts, 171-174, 178, 179
   of neuroblastoma clones, 171-174, 178,
      179
Adrenaline (see Epinephrine)
Affinity chromatography
   of m-RNA, 91, 103
   on oligo-dT cellulose, 91, 103
Aggregated mice
   binding of GABA, 16, 18, 19
   binding of glycine, 16, 18, 19
   brain incorporation of ^{14}C-glucose,
      209
   brain incorporation of ^{14}C-mannose,
      209
   synaptosomal fractions, 16-19
   synaptosomal protein, 16, 17
Aggressiveness
   biochemical concomitants, 113-132
   effect of midbrain lesion, 124
   evoked by radio stimulation, 122, 123
   increased by destruction of
      cingulum, 122
      frontal lobes, 122

251

hippocampus, 122
septal area, 122
ventral midbrain tegmentum, 122
in mice, 113-132
in monkeys, 122, 123
isolation-induced, 113-132
reduction of isolation-induced
by 5-HTP, 115
by lysergide, 115
by monoamine oxidase inhibitors,
115
increase of brain 5-HT, 115
role of brain monoamines, 115-125
role of fornix, 123
spontaneous, inhibited by radio stimulation, 123
Alanine
-alanine collected from amygdala, 3
content and collectability in caudate
nucleus, 12, 13
content in control and thyroidectomized
rats, 68
Alcohol (see Ethanol)
Alcoholism
behavioral changes, 166
in United States, 166
Aminobutyric acid (GABA)
binding to synaptosomal fractions,
effects of environment, 16, 18, 19
content in brain of hypothyroid rat,
66, 68
content, synthesis and collectability in
caudate nucleus, 12, 13
effect on redox potential, 161-163
release, 12, 13, 192
d-Amphetamine
action on B9 cell bodies, 245
decreases tryptophan hydroxylase in
brain, 233, 235, 236
effect on 5-HT system of brain, 225,
233, 235, 236, 245
interaction with lithium, 225, 235, 236,
243, 245
Amygdala
bilateral lesions, taming effect, 121
collection of glutamine, asparagine,
serine, glutamate, glycine, -alanine, 3
collection of labelled aspartate and
citrulline, 3
lowered 5-HT content after olfactory
bulbectomy, 116

perfusion with synthetic spinal fluid, 3
perfusion with U-^{14}C-D-glucose, 3
synthesis of dopamine and noradrenaline,
3
Anterior raphé complex (Rh)
and rhythmic slow waves, 133, 141
"A-type" cells, 139
"B-type" cells, 142
lesion of, 141
Antibody
antibody-like globulins, 215
as chemoreciprocals, 199
for Tay-Sachs disease, 211
reactions with antigens, 204
to astrocytin, 213-215
to malignin, 213-215
Antigen
antigen-like fragments, 211-213
Antiserum
anti-Tay-Sachs, 213
goat, fluorescent, 213
Asparagine
collected from amygdala, 3
content, synthesis and collectability
in caudate nucleus, 12, 13
Aspartate
content, synthesis and collectability
from caudate nucleus, 12, 13
decrease in brain of hypothyroid rats,
66, 68
Astroblast
clonal line NN, 169, 171-179
culture of, 169, 170
effect of ethanol, 169, 171-179
high-affinity choline uptake in, 169, 170
Mg^{++}-ATP-ase, effects of ethanol, 171-174
Na^+, K^+-ATP-ase, effects of ethanol,
171-174
5′-nucleotidase, effects of ethanol, 174-
179
of hamster, 169
Astrocytin
a recognin, 211-213, 220
an antigen-like fragment, 211
antibodies to, 213-217
immunochemical specificity, 219
relation to brain tumors, 211-213
relation (structural) to respiratory
proteins, 217
structure of, 213, 214
synthesis of, 215

"A-type" cell
    as dopaminergic cells, 146
    definition of, 138
    greenish fluorescence, 144
    in VTA, STN and Rh, 136-149
    role in rhythmic waves, 136-149
Automated amino acid analysis
    of amino acids of caudate nucleus, 3,
        12, 13
Axoplasmic flow
    of proteins and glycoproteins to nerve
        endings, 16, 20, 225
    role in memory, 200

B

Basal ganglia
    effects of ECS, 34, 35
    effects of ECS and uric acid on
        glial protein synthesis, 46-49
        glial 5-HT uptake, 44-46
        neuronal 5-HT uptake, 44, 45
        neuronal protein synthesis, 46-49
        synaptosomal fraction, 36
        synaptosomal protein synthesis, 34-36
Binding
    binding sites for GABA and glycine, 16
    of GABA and glycine to synaptosomal
        fractions, effects of environment,
        16, 18, 19
Body weight
    decrease in hypothyroid rats, 66, 67
    effect of L-thyroxine, 71
Bound ribosomes
    effects of light-deprivation, 85-107
    effects of light of different wavelengths,
        102-106
    incorporation of ^{14}C-uridine, 96-103
    incorporation of ^{3}H-phenylalanine,
        94-96, 98-100, 102, 103
    m-RNA, 102-104
    RNA of glia of visual cortex after light
        deprivation, 92
    RNA of neurones of visual cortex after
        light deprivation, 92
Brainstem
    monoamines, 133-152
    relation to behavior, 133-152
    rhythmic slow waves, 133-152
Brain weight

    decrease in hypothyroid rats, 66, 67
    effect of L-thyroxine, 71
Brightness discrimination, 87
"B-type" cell
    as serotoninergic cells, 146, 149
    definition of, 138
    in VTA, STN and Rh, 136-149
    produces membrane hyperpolarization,
        149
    role in rhythmic waves, 136-149
    yellowish fluorescence, 144
Butyric acid
    binding of lecithin, 49
    coma, 54
    complexification of neurotransmitters,
        49
effects
    of intraperitoneal injection, 50-52
    on brain 5-HT, 52-54, 57, 58
    on cerebral protein synthesis, 54-58
    on conditioned avoidance response,
        50-52
    implication in hepatic coma, 49, 50
    interaction with uric acid
        on brain 5-HT content, 57, 58
        on brain protein synthesis, 57, 58
        on retrograde amnesia, 56, 57
    myoclonus, 54
    retrograde amnesic effects, 49-57

C

Cancer
    astrocytin, 211-217, 220
    definition, 198
    malignin, 211-217, 220
    malignant brain tumors, 211-217
    role of recognins, 211-217, 220
    serum TAG test, 212-217, 220
Carbachol
    induces circling movements, 151
    injection, intracerebral, 151
Carbohydrate (see also Glycoprotein and
    Glucose)
    as component of glycoproteins, 203
    as component of mucoids, 203
    galactose, 205-207
    in cell recognition, 203-211
    in "sign-post" theory, 205-211
    mannose, 209

of pigeon brain glycoproteins, 208-210
protein-bound hexose, 210, 211
Catalase
  synthesis by ribosomes, 87
Caudate nucleus
  circling movements, 150
  collection of amino acids, 12, 13
  content and synthesis of amino acids,
    12, 13
  intermittently-programmed stimulation,
    123
  perfusion with U-^{14}C-D-glucose, 12, 13
  radio stimulation, 123
  studied with chemitrode, 3
  synthesis of noradrenaline, 3
  synthesis of dopamine, 3
Cell-cell relationships (see Recognins)
Cerebellar cortex
  effects of ECS, 34, 35
  synaptosomal fraction, 36
  synaptosomal protein synthesis, 34-36
Cerebellum
  action of pentylenetetrazol, 185-192
  action of prostaglandins, 184-193
  and chlordiazepoxide, 188-190, 192
  and convulsions, 185-193
  and harmaline, 189, 193
  cerebral concentration of cyclic-AMP,
    185-190, 192
  cerebral concentration of cyclic-GMP,
    184-193
  climbing fibers, 191, 192
  content of monoamines, 117
  effects of ECS and uric acid on
    glial 5-HT uptake, 44, 46
    glial protein synthesis, 46-49
    neuronal 5-HT uptake, 44, 45
    neuronal protein synthesis, 46-49
  Purkinje cells, 189, 191, 192
  rat, 184-193
  tremors, 184, 189, 191-193
Cerebral cortex
  auditory area, 98
  effects of ECS, 34, 35
    and uric acid on
      glial 5-HT uptake, 44-46
      glial protein synthesis, 46-49
      neuronal 5-HT uptake, 44, 45
      neuronal protein synthesis, 46-49
  frontal area, 98
  polysomes and messenger-RNA, 85-111

protein synthesis, 32-35, 85-111
striate area, 224, 228, 230, 233-235,
  237, 239, 242
synaptosomal fraction, 36
synaptosomal protein synthesis, 34, 35
visual area, 85-111
Cerebrospinal fluid
  action of lithium on 5-HIAA, 243
  glycoproteins, 205
  increase in content of 5-HIAA with
    hepatic coma, 50
  metabolite release blocked by
    probenecid, 243
Chemical stimulation
  in completely free animals, 8
Chemitrode
  collection of amino acids from pre-
    cursor U-^{14}C-D-glucose, 11-14
  dopamine and norepinephrine synthesis
    in caudate nucleus, hypothalamus
    and amygdala, 3
  general use, 1-3, 11-14
  use of tyrosine and DOPA, 3
Chemoreciprocals
  antibody-like structure, 213-219
  in serum diagnostic test, 215-217, 220
  relation to recognins, 197-222
Chlordiazepoxide
  action on brain cyclic-AMP, 188-190,
    192
  action on brain cyclic-GMP, 188-190,
    192
  action on pentylenetetrazol seizures,
    188-190, 192
  as anticonvulsant, 188-190
  comparison with PGE$_2$ as anticonvul-
    sant, 188-192
p-Chloroamphetamine
  specification on B9 cell bodies, 245
p-Chlorophenylalanine
  increases irritability in rats, 116
Choline
  as component of phospholipids, 168
  high-affinity uptake, 169, 170, 178
  low-affinity uptake, 169
  uptake by glial and neuronal cultures,
    169, 170, 178
Circling movements
  and cholinergic substances, 150, 151
  block by tactile stimulation, 149, 152
  by injections into VTA, 144-146

by intracerebral injection, 144-146, 150, 151
by tonic electrical stimulation, 151
contralateral, 145,
in relation to monoamines, 144-146, 150, 151
involvement of
caudate nucleus, 150
hypothalmus, 150
red nucleus, 150, 151
substantia nigra, 145-146, 150
via increase of 5-HT, 151
Climbing fibers
activation of Purkinje cells, 191, 192
synapse with Purkinje cells, 191
Cocaine
action on high-affinity uptake of tryptophan, 236, 238-242
action on tryptophan hydroxylase, 236, 238-242, 245
interaction with lithium, 225, 238-242, 245, 246
Coding molecules
chemoreciprocals, 197-222
glycoproteins, 205-211
recognins, 197-222
requirements for, 201, 203
Colliculus, 86
Compartmentation of brain
metabolic, 14
subcellular, 14
Concanavalin A
binding of lectin, 176
ethanol as a non-competitive antagonist, 176, 178
inhibits ecto-5'-nucleotidase activity, 176
Conformational changes
of macromolecules during experience, 14
via prolonged drug exposure, 171, 174
Convulsion
action on cyclic-AMP, 185-190, 192
action on cyclic-GMP, 185-192
and cerebellar cyclic nucleotides, 183-195
and prostaglandin E_2, 183-193
by electroconvulsive shock, 184
by pentylenetetrazol, 184-192
by picrotoxin, 184
by strychnine, 184

in rat, 184-193
protection by chlordiazepoxide, 188-190
Corpora quadrigemina
decreased 5-HT by isolation, 116-118
Cretinism
cretinous hypothyroid rats, 64
experimental model, 63
injection of ^{131}I, 64
Culture
action of Concanavalin A, 176, 178
astroblasts, 167, 169, 171, 174, 178
ATP-ase, 170-174
choline uptake, 168-170
effects of ethanol, 165-179
glioblasts, 169, 174, 178
glioma, 174, 178
membrane-associated enzymes, 170-179
membrane topography, 167-168
neuroblastoma, 169, 171, 174, 178
5'-nucleotidase, 174-179
of neural cells, 165-174
Cyclic-3', 5'-adenosine monophosphate (cyclic-AMP)
assay, 185
effect of chlordiazepoxide, 188-190, 192
effect of pentylenetetrazol, 185-192
in cerebellum, 185-190, 192
interaction with prostaglandins, 184, 188, 192
Cyclic-3', 5'-guanosine monophosphate (cyclic-GMP)
and convulsions, 184-193
and tremors, 184, 189, 192, 193
assay, 185
decrease in cerebellum, 191
effect of
chlordiazepoxide, 188-190, 192
harmaline, 189, 193
pentylenetetrazole, 185-192
prostaglandins, 184-192
in cerebellum, 185-193
increase by glutamate, 192
relation to Purkinje cells, 191
2', 3'-Cyclic nucleotide-3'-phosphohydrolase (CNP)
decrease in brain of hypothyroid rat, 66, 70, 71, 73
effect of
bovine somatotropin, 71, 73

hydrocortisone, 73
L-thyroxine, 71, 73
marker for myelin sheath, 66
method of determination, 64
relation to learning ability, 71, 74, 75
5'-Cytidine monophosphate (5'-CMP)
as substrate for ecto-5'-nucleotidase, 176, 177
as substrate for total 5'-nucleotidase, 176, 177
effect of ethanol, 176, 177
Cytochrome b_5
human, 217

D

Dark-rearing
effect on
base composition of m-RNA, 102-104
^{14}C-uridine incorporation into RNA, 100-102, 106, 107
^{14}C-uridine incorporation into RNP, 96-99
^{3}H-phenylalanine incorporation into polyphenylalanine, 100
^{3}H-phenylalanine incorporation into protein, 99, 100, 106, 107
^{3}H-phenylalanine incorporation into RNP, 94-96, 98-100
RNA distribution in visual cortex, 92-94
of rats, 92-107
Darwin, C., 15
Deiters' nucleus
increase in nuclear RNA of neurones and glia during training, 15
Delivery systems
pharmacological, 212
Dendrites, 86
Dendritic spines, 86
Density gradient fractionation
for separation of neurones from glia, 90
of synaptosomal-mitochondrial fractions of mouse brain, 16-19, 34, 35
Deoxyribonucleic acid (DNA)
brain content in hypothyroid rats, 66, 67
brain content in phenylketonuric rats, 73

decrease in MAM-treated rat brain, 77
decrease in microencephalic brain, 77
effect of growth hormone, 74
effect of hydrocortisone on brain metabolism, 73
method of determination, 92
relationship with mucoid-contact functions, 210, 211, 220
synthesis in developing rat brain, 81-83
synthesis stimulated in tumors, 211
Dialytrode, 1-3
Diazepam
in study of individual rank in monkeys, 6
protection against tremor of harmaline, 185-191
Diencephalon
decreased 5-HT by isolation, 116-118
effects of ECS, 34, 35
and uric acid
on glial 5-HT uptake, 44-46
on glial protein synthesis, 46-49
on neuronal 5-HT uptake, 44, 45
on neuronal protein synthesis, 46-49
synaptosomal fraction, 36
protein synthesis, 34-36
Dihydroxyphenylalanine (see DOPA)
5, 6-Dihydroxytryptamine
degeneration of cerebral 5-HT terminals, 119
production of rat muricidal behavior, 119
DNA-dependent DNA polymerase
method of determination, 64
peak activity at sixteenth post-natal day in rat brain, 81
DOPA (dihydroxyphenylalanine)
as precursor for dopamine and norepinephrine, 3
use with chemitrode, 3
Distance factor
from brain tumor, 210
Dopamine
content of STN and VTA, 143-145
dopaminergic neurones, 146, 149, 152
fluorescence staining, 144, 152
increase in brain
of hypothyroid rat, 66, 69
of microencephalic rat, 77, 79
regions by Nembutal, 143, 148

increased turnover in brains of isolated
mice, 118
in relation to rhythmic slow waves,
143-145
method of determination, 135
studied with chemitrode, 3
synthesis in brain regions, 3

E

ECS (see Electroconvulsive shock)
Ectoenzyme
in relation to ethanol, 167, 174-178
5'-nucleotidase, 174-178
Electrical stimulation (see also Electroconvulsive shock)
extensor seizures, induction of, 184
in completely-free animals, 8
Electroconvulsive shock (ECS)
age-specificity of effects, 26, 31-33
change in brain 5-HT content, 25, 28-30
effect
of prostaglandin, 184
of uric acid, 27-35, 37, 38, 42-49,
58-59
on brain 5-HIAA, 28
on cellular 5-HT uptake, 42-46
on protein synthesis, 26, 31-35,
46-49, 58, 59
on synaptosomal 5-HT uptake, 37,
38
induction of
extensor seizures, 184
retrograde amnesia, 25-59
transcorneal application, 30, 32, 34, 42
Electron microscopy
of synapse, 210
Endoplasmic reticulum
incorporation of fatty acids, decrease
by light deprivation, 106
Enriched environment
effect on glial proliferation, 88
Environmental factors
as cause of learning, memory and
behavior, 11
as "trigger" of behavioral changes, 11,
114-125
brain growth through experience, 15
effect on
cerebral morphology, 14-16

GABA and glycine binding in brain
synaptosomes, 16-19
protein content of synaptosomal
fractions, 16-17
environmental stimulation, 15
exposure to light of different wavelengths, 85-111
in neurochemistry of behavior, 11-19
in neurochemistry of individuality, 4
in relation to aggressive behavior, 16,
113-132
isolation-induced aggressiveness in
mice, 16
light deprivation, 85-111
on cerebral development, 85-111
Enzyme
ATP-ase, 170-174, 178
ecto-5'-nucleotidase, 174-179
ectoenzyme, 167, 174, 179
effects of ethanol, 170, 179
galactose transferase, 205
in cell cultures, 170-179
neuraminidase, 167-169
5'-nucleotidase, 174-179
tryptophan hydroxylase, 223-246
Epinephrine
effect on redox potential, 161-163
Eserine
effect on redox potential, 158, 159
Ethanol
acute exposures of cell cultures, 167-
179
antagonism of effects by lithium 245,
246
as non-competitive antagonist of Concanavalin A, 176, 178
chronic exposures of cell cultures, 167-
179
effects on
cell cultures, 165-182
central nervous system cells, 165-
182
choline uptake, 168-170
membrane-associated enzymes,
170-179
membrane topography, 167-169
on ATP-ase, 170-174, 178, 179
on 5'-nucleotidase, 174-179
Evoked potential
and rhythmic waves, 135, 136
between STN and VTA, 135, 136

negative peak, 139, 141
positive phase, 141
Excitation
repeated excitation of hippocampus, 4

F

Fatty acid
incorporation into endoplasmic
reticulum, 106
Ferredoxin
of alfalfa, 217
of leucaene glauca, 217
Ferritin
synthesis on free polysomes, 87
Fluorescence histochemistry
of 5-HT, DA and NE in VTA and STN
after harmaline, 145
after 5-HTP, 145-147
after Nembutal, 145
method for monoaminergic neurones,
134, 145-149, 151, 152
Fluorography, 91
Free ribosomes
effects of light deprivation, 85-107
effects of light of different wavelengths, 102-106
incorporation of ^{14}C-uridine, 96-103
incorporation of ^{3}H-phenylalanine,
94-96, 98-100, 102, 103
m-RNA, 102-104
RNA of glia, visual cortex with light
deprivation, 92
RNA of neurones, visual cortex with
light deprivation, 92

G

Galactose transferase
action, 205
in "sign-post" theory, 205
role in intercellular contact, 205
Gel electrophoresis
of synaptic protein, 16
Generating mechanism
of rhythmic slow wave, 134, 147-149,
151, 152
Genetic factors
in neurochemistry of behavior, 12, 14

in neurochemistry of individuality, 4
Glia
alteration in phenylketonuria, 73
astrocytes, 41, 44
cancer, 212, 213, 215-220
decreased RNA/DNA ratio of visual
cortex, 92
effects of light-deprivation, 85-111
functional reciprocity with neurones,
48
glial cultures, effects of ethanol
astroblast, 169-179
glioblast, 169, 170, 174-179
glycoprotein 10B as glial constituent,
15, 16
increased protein synthesis by ECS,
48, 49
nuclear RNA, 15
oligodendrocytes, 41, 43, 73
polysomes and messenger-RNA, 85-111
possible effect of growth hormone, 75,
76
proliferation in neocortex upon exposure
of rats to enriched environments, 88
protein synthesis, ECS and uric acid
effects, 46-49
role in information processing, 15, 16
S-100 and 14-3-2 proteins as glial constituents, 15, 16
separation from neuronal perikarya,
38-44, 90
tumors, 210-213, 215-220
uptake of 5-HT, effects of ECS and
uric acid, 42-46
Glioblast
culture of, 169, 170
effect of ethanol, 169, 174-179
high-affinity choline uptake in, 169,
170
inhibition of ecto-5'-nucleotidase by
Concanavalin A, 176
5'-nucleotidase activity, 174-178
of rat, 169, 174-179
Glioma
astrocytin, 211-215
ependymoma-glioma, 210
malignant, brain, 211-217
malignin, 211-215
recognins, 211-215
Globin
synthesis by ribosomes, 87

Glucose
  as precursor of glycoproteins, 209, 210
  glucose incorporation into training and
      nesting pigeon brain glycoproteins,
      209, 210
  incorporation of ^{14}C-glucose into
      brains of isolated and aggregated
      mice, 209
  U-^{14}C-D-glucose as amino acid precursor
      in brain, 3, 11-14
Glutamate
  as excitatory transmitter, 192
  as substrate for mitochondria, 156
  content synthesis and collectability
      in caudate nucleus, 12, 13
  decrease in brain of hypothyroid rat,
      66, 68
  glutamate seizure, 3
  L-glutamate use with dialytrode, 2, 3
Glutamine
  collected from amygdala, 3
  content, synthesis and collectability in
      caudate nucleus, 12, 13
Glycine
  "binding" to synaptosomal fractions,
      effects of environment, 16, 18, 19
  collected from amygdala, 3
  content, synthesis and collectability in
      caudate nucleus, 12, 13
Glycoprotein
  action of neuraminidase, 168, 169
  as component of Na$^+$, K$^+$-ATP-ase, 174
  axoplasmic flow, 16, 20
  brain, 15, 16, 20, 87, 167-169, 174, 198,
      202-211, 219, 220
  glycoprotein
      10B as glial constituent, 16, 211
      10B as membrane constituent, 16
      11A, 208, 209
      groups, 208, 209
  high concentration at synapses, 204,
      210
  incorporation of ^{14}C-glucose, 200, 209
  role in cell recognition, 198, 202-211
  sialic acid, 167-169
  "sign-post" theory, 198, 204-210
  synthesis enhanced during learning or
      training, 15, 207-210
  synthesis on bound ribosomes, 87
Growth hormone
  effect on

brain DNA, 74
brain weight, 74
myelination, 74, 75
post-natal brain development, 74-76
experimental model of excessive
    secretion, 63
increase in
    body weight, 74
    cerebral CNP activity, 74
    learning ability, 74-76
injection of bovine somatotropin, 64
possible effect on glia, 75, 76
Growth of brain
    cellular, 14-17
    "growth through experience," 14-17
    molecular, 14, 15
5'-Guanosine monophosphate (5'-GMP)
    as substrate for exto-5'-nucleotidase,
        176, 177
    as substrate for total 5'-nucleotidase,
        176, 177
    effect of ethanol, 176, 177

H

Harmaline
    action on "B-type" cells, 147
    activation of olivocerebellar system,
        189, 193
    effect on cerebral monoamines, 142-
        146
    enhancement of postural tremor, 150
    enhances Purkinje cell firing, 189
    increases cerebellar cyclic-GMP, 184,
        189, 191-193
    induction of
        cerebellar tremors, 184, 189, 191-
            193
        circling movements, 145, 150
        rhythmic slow waves, 136, 137, 142,
            143, 152
    inhibitor of 5-HT oxidase, 135, 149,
        150
    interaction with prostaglandins, 189,
        191-193
    intracerebral injection, 144-146
Hebb, D.O., 14
High-affinity system
    for choline uptake, 169-170
        effect of ethanol, 169-170

for 5-HT by synaptosomes, 224
for tryptophan uptake, 223, 224, 234, 236
    effect of cocaine, 236, 239
    effect of lithium, 234
Hippocampus
    acid phosphatase in regions, 4,5
    cholinesterase in regions, 4, 5
    dorsal area, 4, 5
    increased aggressiveness with lesions, 122
    increases in ribosomesomes after acquisition of brightness discrimination or conditioned reflex, 87
    involved in integration of nociceptive responses, 123
    repeated excitation, 4
HSS (High-speed supernatant)
    content of RNA in liver, 92
    content of RNA in visual cortex, 92
    effects of light-deprivation, 94-99
    incorporation of ^{3}H-phenylalanine, 95
    method of preparation, 90
Hydrocortisone
    decrease in CNP activity, 73
    effect on cerebral DNA, 73
    effect on learning ability, 73
6-Hydroxydopamine
    depletion of brain catecholamines, 119
    facilitates shock-elicited aggression, 119
5-Hydroxyindoleacetic acid (5-HIAA)
    decrease with monoamine oxidase inhibition, 28
    increase
        in brain regions with hepatic coma, 49
        in CSF with hepatic coma, 50
        with retrograde amnesia, 28
    of cerebrospinal fluid, 243
    turnover affected by lithium, 243
5-Hydroxytryptamine (Serotonin)
    butyric acid-uric acid interactions on brain content, 57, 58
    content in VTA and STN, 143-146
    decreased turnover in isolated mice, 116, 118-125
    developmental changes, brain, 119
    effect of
        butyric acid, 52-54
        ECS on synaptosomal uptake, 37-39
        electroconvulsive shock, 26, 28-30

harmaline of decreasing its degradation, 150
rhythmic slow waves on content, 142-146
uric acid on synaptosomal uptake, 37-39
effects on protein synthesis, 26, 29, 30, 32
elevation with training, 25, 26
implication in hepatic coma, 49, 50
increase in brain of hypothyroid rat, 66, 69
increase in brain of microencephalic rat, 77, 79
increased content with Nembutal, 143, 148
induction of circling movements, 150
induction of tremor, 150
injection into midbrain raphé, 151
injection into substantia nigra, 150
intracranial administration, 25
intrahippocampal administration, 30
involvement in
    isolation-induced aggressiveness in mice, 115-125
    manic-depression, 225, 243, 246
    retrograde amnesia, 25-59
modified turnover with ECS, 28
nerve endings, 223
no change in turnover during isolation, 118-119
regions of brain, turnover in, 118
relation to behavior, 144-146, 243
release from nerve endings, 244
serotoninergic neurones, 146
sex differences in turnover, 120
synaptosomal uptake, 37-38
synthesis in brain, 223-225, 229, 231-233, 235-246
turnover in different mouse strains, 121
uptake, 224, 228
    by neurones and glia
        effects of ECS, 42-46
        effects of uric acid, 42-46
5-Hydroxytryptophan (5-HTP)
    effect on brain monoamine content, 142-144
    effect on "B-type" cells, 147, 148
    increase in brain 5-HT, 115
    induction of circling movements 145, 150
    induction of rhythmic slow waves, 136,

137, 142-146, 148, 150, 152
intracerebral injection, 135, 144, 145, 150
intravenous injection, 136, 137, 142
Hypothalumus
circling movements, 150
lateral, 121
lesion of ventro-medial region, 122
medial, 121
role in aggressiveness, 121
role in feeding behavior, 121
studied with chemitrode, 3
synthesis of dopamine, 3
synthesis of norepinephrine, 3

I

Immunofluorescence
of brain tumor, 216
serum TAG test, 216
to label glial cancer cells, 213, 216
Immunoglobulin
synthesis on bound ribosomes, 87
Information processing, 198, 201-210, 219, 220
5'-Inosine monophosphate (5'-IMP)
as substrate for ecto-5'-nucleotidase, 176, 177
as substrate for total 5'-nucleotidase, 176, 177
effect of ethanol, 176, 177
Instrumental behavior, 6-7
operant behavior techniques, criticism, 6-7
Isolated mice
aggressiveness, 16, 115-125
binding of GABA, 16, 18, 19
binding of glycine, 16, 18, 19
brain incorporation of ^{14}C-glucose, 209
brain incorporation of ^{14}C-mannose, 209
synaptosomal fractions, 16-19
synaptosomal protein, 16, 17
Isolation (environmental)
aggressive "isolated" mice, 16-19, 113-132
biochemical concomitants, 16-19, 113-132
effect on 5-HT turnover, 116-125
increase in cerebral dopamine turnover,

118, 119
increased tyrosine hydroxylase of rat brain, 118
reduction in cerebral norepinephrine turnover, 118, 119
Isoleucine
content of caudate nucleus, 12, 13

K

Ketamine
as anaesthetic in cats, 135
Kinetics of tryptophan uptake, 228-245
Krech, D., 15

L

Lateral geniculate nucleus, 86
Learning
as a biochemical process, 201, 219, 220
as a component of behavior, 11
cerebral incorporation of glucose, 209, 210
coding molecules, 201, 206
effect of MAM-induced microencephaly, 80
glycoprotein 11A, 207-209
involvement of
brain glycoproteins, 205-210
cerebral protein, 14-16, 207-210
cerebral RNA, 15
learning
ability of hypothyroid rats, 66-72
ability-neurochemical correlates, 63-84
test based on "correct response ratio," 65, 66, 72, 75, 76, 80
relation to cerebral morphology, 14-16
"sign-post" theory, 198, 205-210
Leucine
content of caudate nucleus, 12, 13
intraventricular ^{14}C-leucine, 54
use of ^{14}C-leucine for studying protein synthesis, 32-35, 46-49, 54-58
Light-deprivation
decrease of fatty acid incorporation into brain endoplasmic reticulum, 106
effect on polysomes, 85

via enucleation, 86
via rearing animals in a dark enclosure,
85
Light-exposure
different wavelengths, 85
effect on polysomes, 85
Lithium
action on brain 5-HT mechanisms,
224-246
antagonism of ethanol and morphine,
245, 246
biphasic effects, 243
effect on
macromolecular mechanisms, 225
neurotransmission, 225
tryptophan hydroxylase, 224, 229,
231-245
tryptophan transport, 224, 228-245
inhibition of catecholamine release, 228
interaction with d-amphetamine, 225,
235, 236, 243, 245
interaction with cocaine, 225, 236, 238-
242, 245, 246
in treatment of manic-depressive
psychosis, 223-225, 240, 242, 243,
246
latency to effect, 240
long-term administration, 224, 225, 228,
231-233, 244
neurobiological theory of action,
223-249
regulatory role in 5-HT transmission,
243
short-term administration, 224, 225,
230-233, 244
tolerance, 242
Liver
unaffected by light-deprivation, 96
Low-affinity system
for choline uptake, 169
Lysergide
increases brain 5-HT, 115

M

Macromolecules
changes during experience, 14-16
conformational changes, 14
effect of lithium, 225
involvement in behavior, 14-17

involvement with retrograde amnesia,
25, 26, 31-35, 46-49, 54-59
relationship with memory consolidation,
30, 58, 59
Malignin
a recognin, 211-213, 220
an antigen-like fragment, 211
antibodies to, 213-217
immunochemical specificity, 219
relation to brain tumors, 211-213
structural relation to respiratory
proteins, 217
structure of, 213, 214
synthesis of, 215
Manic-depressive disease
action of lithium, 223-249
action of tryptophan in humans, 243
bipolar type, 225, 243
neurobiological theory of, 223-249
role of 5-HT, 243
Maturation
as determinant of behavior, 14
DNA synthesis, 81-83
in relation to learning, 63-84
in relation to neurochemistry, 63-84
relation to
cerebral protein synthesis, 31-33
ECS-induced retrograde amnesia,
30-33
effects of uric acid, 30-33
Membrane topography
constituents, 204
effects of ethanol, 167-169, 178
in gial and neuronal cultures, 167-169,
178
membrane interfaces, 204
membrane surfaces, 204
studied with neuraminidase, 167-169,
178
Memory
as a biochemical process, 201, 219,
220
as a component of behavior, 11, 198
coding molecules, 201, 206
glycoprotein 11A, 207-209
involvement of
brain glycoproteins, 205-210
cerebral protein, 14-16, 29, 30, 58,
59, 204-210
cerebral RNA, 14-16
relation to cerebral morphology,

14-16, 198
"sign-post" theory of, 198, 205-210
Mental retardation
  inborn errors of metabolism, 63
Messenger-RNA
  and polysomes, 85-111
  base composition, 91, 102-104
  bound-polysomal, 104
  degradation of non-poly-A portion, 91
  free-polysomal, 104
  glial, 102-104
  isolation method, 91
  neuronal, 102-104
  purification by affinity chromato-
    graphy, 91, 103
  U/C and A/G ratios, 103, 104
Methionine sulfoxide
  presence in caudate nucleus, 12, 13
Methylazoxymethanol (MAM)
  decrease in cerebral DNA, 77
  decreased size of cerebral hemispheres,
    77, 78
  effect on learning, 80
  increase in cerebral dopamine and
    norepinephrine, 77, 79
  increase in cerebral 5-HT, 77, 79
  no change in cerebral amino acids, 77
  production of microencephaly, 64,
    76-80
Microencephaly
  decreased cerebral DNA, 77
  decreased size of cerebral hemispheres,
    77, 78
  effect on learning, 80
  experimental model, 63
  no change in brain amino acids, 77
  production by methylazoxymethanol
    injection, 64, 76-80
  rats, 76-80
Midbrain
  and circling movements, 144, 145, 150,
    151
  content of monoamines, 117
  decreased tryptophan hydroxylase by
    d-amphetamine, 233-236
  effects of ECS, 34, 35
  effect of lithium on soluble enzyme
    (tryptophan hydroxylase) activity,
    234-236
  5-TH turnover, 118
  lateral area, 233, 234

  lesion to raphé region, effect on
    isolation-induced aggressiveness in
    mice, 124
  raphé nucleus, 135-152
  red nucleus, 145, 150, 151
  relation to rhythmic slow waves, 133-
    152
  reticular formation, 145
  serotonin synthesis in
    effect of d-amphetamine, 233-237
    effect of cocaine, 236, 238, 240-
      243
    effect of lithium, 234-236, 242
    effect of morphine, 236, 238-240
Mitochondria
  effect of
    acetylcholine, 157-163
    epinephrine, 161
    GABA, 161-162
  general, 93, 155
  redox potentials, 156-163
Monkey
  amino acid profile of brain, 11-13
  in study of individual rank, 5-6
  sensitivity to diazepam, 6
  synthesis of
    amino acids in brain, 3, 11-13
    dopamine in brain, 3
    norepinephrine in brain, 3
  use of chemitrode in brain regions, 3
Monamine
  and rhythmic slow waves, 133-152
Monamine oxidase
  5-HT oxidase, inhibition by harmaline,
    135, 148, 149
  inhibition by tranylcypromine, 28, 115
  inhibitors, 28, 115
  possible inhibition by Nembutal, 148,
    149
Monoamine oxidase inhibitor (MAOI)
  harmaline, 5-HT oxidase inhibitor, 135,
    148, 149
  possible action of Nembutal, 148, 149
Morphine
  antagonism by lithium, 245, 246
  effect on brain 5-HT synthesis, 236,
    238
Morphology of brain
  changes with environmental stimulation,
    14-17
  effects of exposure to light of different

wavelengths, 85-111
effects of light deprivation, 85-111
neurones and glia, 38-44
regional synaptosomal morphology,
35-37
relation to learning, memory, behavior,
12-19
single astrocyte, 44
single neuronal cell body, 42
single oligodendrocyte, 43
Mouse
"aggregated," 16, 18, 19, 115-125, 209
incorporation of ^{14}C-glucose into
brain, 209
incorporation of ^{14}C-mannose into
brain, 209
"isolated," 16, 18, 19, 115-125, 209
Reeler disease of brain, 198, 219
Mucoid
and brain tumors, 210, 211
and DNA replication, 210, 211, 220
contact functions, 210, 211
definition, 203
function at membrane interfaces, 204
function at membrane surfaces, 204
history of, 203-205
in synaptosomal fractions, 204
of brain, 203-105
protein-bound hexose, 210, 211
relation to antibody reactions, 204
role in "recognition" phenomena,
203-205
Muricidal behavior
block by bilateral lesions of amygdala,
121, 122
in rats, 115, 116
isolation-induced, 116
olfactory bulbectomy-induced, 116,
122
production by 5, 6-dihydroxytrypta-
mine, 119
Myosin
synthesis on free polysomes, 87

N

NADPH cytochrome c reductase
synthesis by ribosomes, 87
Nembutal (see Pentobarbitone)
Nernst equation, 162

Nerve ending (see also Synaptosome)
axoplasmic flow to, 16, 20, 225
of striate cortex, 228, 233, 236, 237
serotoninergic, 223, 224
tryptophan hydroxylase, 223, 224, 229,
230, 233-245
uptake of tryptophan, 223, 224, 228,
230, 234, 236, 238, 240, 242-245
Neural cells
in culture, 165-179
Neuraminidase
action on glycoproteins, 168, 169
action on sialic acid, 167-169
effects of ethanol, 167-169
in hamster astroblasts, 167-169
sialic acid release, 167, 168
studied in membrane topography,
167-169
Neuroblastoma
adrenergic clone M_1, 169
ATP-ase activity, 170-174, 178
cholinergic clone, S_{21}, 169
culture, 169, 170
effect of ethanol, 169-178
high-affinity choline uptake, 169, 170
5'-nucleotidase activity, 174-179
Neurochemistry of individuality, 4-6
environmental factors, 4
genetic factors, 4
individual rank as an important determinant of pharmacological effects, 6
study with dominant and submissive
monkeys, 5
use of diazepam, 6
Neuronal activity
"A-type" cell, 138-149
"B-type" cell, 138-149
unit activity, 136-142, 151, 152
Neuronal cell bodies
decreased
5-HT uptake by ECS, 44, 45
protein synthesis by ECS, 47
RNA/DNA ratio of visual cortex, 92
effect of
ECS, 42-46
light-deprivation, 85-111
uric acid, 42-46
of brain regions, 38-45
preparation, 38-42, 90
protein synthesis, 46-49
scanning electron micrograph of

neuronal cell body, 42
uptake of 5-HT, 42-46
Nissl substance (see Ribonucleoprotein), 87
Norepinephrine (Noradrenaline)
content, in STN and VTA, 142, 143
increase in brain
of hypothyroid rat, 66, 69
of microencephalic rat, 77, 78
with Nembutal, 143, 148
method of determination, 135
reduction of cerebral turnover by isolation, 118
relation to rhythmic slow waves, 142, 143, 148, 152
studied with chemitrode, 3
synthesis in brain regions, 3
Nuclei (of cell), 93
5'-Nucleotidase
as an ectoenzyme, 174
effect of acute ethanol exposure, 174-177, 179
effect of chronic ethanol exposure, 174-177, 179
inhibition by Concanavalin A, 176, 178
studied in cell cultures, 174-179
substrates for, 176, 177

O

Olfactory bulb
monoamine content, 117
Olfactory bulbectomy
eliminates natural killing in rats, 122
induced muricidal behavior, 116, 122
lowered 5-HT of amygdala, 116
reduction of aggressiveness, 122
Olivocerebellar system
activation by harmaline, 189, 193
Operant behavior
and brain glycoproteins, 208-210
criticism of technique, 7
in pigeons, 204, 207-210
operant brightness discrimination, 65, 66
Optic tract, 86
Oxidative phosphorylation
media for measurement, 156
Oxygen consumption
media for measurement, 156

P

Passive avoidance behavior
in mouse, 26-28, 30, 31, 50-52, 56, 57
Pentamethylenetetrazol (Metrazol)
convulsions, 184-190, 192
effect
of PGE_2, 185-192
of prostaglandins, 184-190, 192
on cerebellar cyclic nucleotides, 185-192
Pentobarbitone (Nembutal)
and presynaptic inhibition, 147
as a monoamine oxidase inhibitor, 148, 149
as anaesthetic in cats, 134, 135
production of rhythmic slow waves, 133, 135, 136, 142, 149, 152
production of tremor, 149, 150
Permeability
of membrane, 155
Phenylalanine
as precursor for protein synthesis, 92, 94-103
content of caudate nucleus, 12, 13
incorporation into RNP, 95, 98, 99
L-phenylalanine diet, 64
serum levels, 64
Phenylketonuria
decrease in
body weight, 73
brain 5-HT, 73, 74
brain weight, 73
cerebral CNP activity, 73, 74
experimental model, 63
increase in brain phenylalanine, 73, 74
increase in brain tyrosine, 73, 74
neurochemical data, 73, 74
no change in cerebral DNA, 73
phenylketonuric rats, 73-75
production by high phenylalanine diet, 64
Phenylpyruvic acid
urinary excretion, 64
Phospholipid
decrease in brain of hypothyroid rat, 66, 70
increase in brain during myelination, 66
involvement of choline in synthesis, 168
Picrotoxin
effect of prostaglandins, 184

induced convulsions, 184
Pigeon
  brain glycoproteins, 200, 207-210
  carbohydrate constituents, 207-210
  incorporation of glucose into brain
    glycoproteins, 200, 209, 210
  learning and memory, 207-210
Plasticity
  chemical and anatomical, 14, 113
  of brain, 113, 155
  of neurones, 155
  relation to learning, memory and
    behavior, 14
Polyphenylalanine
  as a measure of monomer to polysomal
    ratio, 99
  incorporation of ^{3}H-phenylalanine, 99,
    100
Polysome (see also Ribosome)
  and messenger-RNA, 85-111
  free, 85
  membrane-bound, 85
  neuronal and glial, 85-111
Proline
  content in caudate nucleus, 12, 13
Prostaglandin
  and cyclic nucleotides, 183-195
  and convulsive drugs, 183-195
  and Purkinje cells, 184
  and tremorogenic drugs, 183-195
  direct action on central neurones, 184
  effect on cerebellar cyclic-GMP,
    189-193
  effects on convulsants, 184, 188, 189
  microelectrophoretic application, 184
  prostaglandin E_2, 183-193
  protection against harmaline-induced
    tremors, 189
  protection against pentylenetetrazol-
    induced seizures, 184, 188, 189
  release from CNS regions, 183
  synthesis in brain, 183
Protein (see also Enzymes, Glycoprotein)
  age difference in synthesis, 31-33
  amine regulation in retrograde amnesia,
    25-59
  axoplasmic flow, 16, 20
  cellular synthesis, effects of ECS and
    uric acid, 46-49
  changes involved in experience, 14, 16,
    17, 34, 35

content of synaptosomal fractions as
    altered by environment, 16, 17
  effects of butyric acid on synthesis,
    54-58
  effects of uric acid on synthesis, 31-35,
    46-49, 57-59
  in brain regions, 34, 35, 46-49, 54-56,
    57, 58
  14-3-2 protein
    as glial constituent, 15, 16
    as membrane constituent, 16
    enhanced during learning and train-
      ing, 15
    preferential synthesis on free ribo-
      somes, 87, 90, 91
  S-100 protein
    as glial constituent, 15, 16
    as membrane constituent, 16
    enhanced during learning or train-
      ing, 15
    preferential synthesis on free ribo-
      somes, 87, 90, 91
  soluble, of HSS fraction, 94
  synaptosomal synthesis, effects of ECS
    and uric acid, 34, 35
  synthesis
    enhanced by learning, 15
    reduced by 5-HT, 26
    studied with ^{14}C-leucine, 32-35,
      46-49, 54-58
Pteridine cofactor
  use with assay for tryptophan hydro-
    xylase, 225
Purkinje cells
  action of harmaline, 189
  action of prostaglandins, 192
  genetic degeneration of, 191
  location of prostaglandin dehydrogenase,
    192
  relation with cyclic-GMP, 191
  synapse with climbing fibers, 191, 192
Puromycin
  effect on training, 15
Push-pull cannula, 1-3

R

Radio-stimulation
  use in completely-free animals, 7-8
Ramón y Cajal, S. 14, 15
Raphé (see also Midbrain)

axonal flow, 232
cell bodies, 223, 224, 226, 232, 233,
    240, 244, 245
effect of
    d-amphetamine, 233, 235-237, 245
    cocaine, 236-245
    morphine, 238
    tryptophan hydroxylase, 223, 224,
        232, 233, 236, 240-246
Receptor
    viral, 204
Recognins
    and their chemoreciprocals, 197-222
    astrocytin, 211-220
    definition of, 203, 219
    in diagnosis of brain malignancy, 211
        219
    malignin, 211-220
    precursors, 199
Recognition substances (see Recognins)
Red nucleus
    and circling movements, 150, 151
    lesion, 151
    microinjection of harmaline, 145
    microinjection of 5-HTP, 145
Redox potentials
    apparatus for measurement, 156, 157
    effect of
        ACh, 158-163
        epinephrine, 161-163
        eserine, 159
        GABA, 161-163
    of mitochondria, 156-163
    with succinate, 156, 158-162
Reeler disease
    genetic error, 198
    in mouse, 198
    recognin, 219-220
Regional neurochemistry
    acid phosphatase in regions of hippo-
        campus, 4-5
    cholinesterase in regions of hippo-
        campus, 4-5
    criticism of studies with homogenates,
        3,4
Release
    of amino acids from brain, 2-3
    studied with
        chemitrode, 1-3
        dialytrode, 1-3
        push-pull cannula, 1-3

Retina, 86
    proteins, effect of light exposure, 104
Retrograde amnesia
    age-related differences, 30-34
    amine regulation of protein synthesis,
        25-59
    butyric acid-induced, 56, 57
    effects of butyric acid, 49-52, 56-58
    effects of uric acid, 26-35, 37, 38,
        42-49, 56-59
    production by ECS, 25-59
    relation to
        brain 5-HT, 25, 26, 28-32
        cellular protein synthesis, 46-49
        neuronal and glial 5-HT uptake,
            42-46
        oxidative metabolism, 58, 59
        synaptosomal 5-HT uptake, 37-39
Rhythmic slow waves
    and evoked potentials, 135, 136
    and unit activity, 136-142, 152
    generating mechanism, 134, 147-149,
        151, 152
    in anterior raphé complex, 133-135,
        139-142, 152
    in brainstem, 133-154
    in cats, 133-152
    inhibition by pallidal stimulation, 133
    in regulation of involuntary movement,
        133-134, 146
    in sleep-waking cycle, 134, 151
    in subthalamic nucleus, 133-152
    in ventral tegmental area, 133-152
    produced by
        intracerebral injection
            of harmaline, 144-146
            of 5-HTP, 144-146
        i.v. injection
            of harmaline, 136, 137, 142-144
            of 5-HTP, 136, 137, 142-144
        Nembutal, 142, 143
    relation
        to barbiturate tremor, 149, 150
        to circling movements, 145, 146,
            148-152
        with behavior, 133-154
        with monoamines, 133-154
Ribonucleic acid (RNA)
    distribution in subcellular fractions of
        visual cortex of dark-reared and
        light-exposed rats, 92-95

involvement in learning and memory, 15
messenger- ,
    base composition, 91
    isolation, 91
    method of determination, 92
    pH6-RNA, 100-103
    pH8-RNA, 101-103
    ribosomal- , isolation, 91
    RNA base ratios, 15
    RNA/DNA ratio, 92
Ribonucleoprotein
    incorporation of ^{3}H-phenylalanine,
        95, 98, 99
    incorporation of ^{14}C-uridine, 97-99
    of HSS fraction of visual cortex, 95,
        97, 99
    of ribosomal fraction of visual cortex,
        95, 97, 99
Ribosomes (see also Polysomes)
    effect of dark rearing with subsequent
        light exposure, 94
    free, 87
    free and bound polysomes, 85-111
    membrane-bound, 87
    membrane-ribosome interaction, 94
    synthesis of proteins, 87
    transcription of RNA, 100
RNA/DNA ratio
    decrease in neurones and glia of visual
        cortex with light deprivation, 92, 93

S

S-100 Protein (see Protein)
    synthesis with response to flickering
        light, 87, 104
Scanning electron microscopy
    of individual synaptosome, 36, 37
    of neurones and glia, 36-44
Serine
    collected from amygdala, 3
    content, synthesis and collectability in
        caudate nucleus, 12, 13
Serum albumin
    synthesis on bound ribosomes, 87
Sherrington, C.S., 14
Sialic acid
    as component of glycoproteins, 167-169
    as component of sialolipids, 167-169
    distribution in astroblasts, effect of
        ethanol, 168, 169, 178

    in membrane topography, 167-169,
        178
    release by neuraminidase, 167, 168, 178
Sialolipid, 167-169
"Sign-post" theory
    of brain circuitry, 198, 205-210, 219,
        220
    of learning and memory, 198, 205-210,
        219, 220
Sleeping
    rhythmic slow waves, 134, 151
    sleep-wakefulness cycle, 134, 151
Somatotropin (see Growth hormone)
Spontaneous (free) behavior
    field studies, 7-8
    social behavior of free animals, 7-8
    use of gibbons, 7, 8
Spurzheim, J.G., 15
Strychnine
    effect of prostaglandin, 184
    induced convulsions, 184
Subcellular fractionation, 16-19, 34-59,
    89, 90, 91, 156, 226
Substantia nigra
    and circling movements, 145, 150, 151
    and tremor, 150
Subthalamic nucleus (STN)
    and rhythmic slow waves, 133, 135-145,
        147-152
    connection with VTA, 144-147
    effect of raphé lesions, 151
    evoked potentials, 135-136
    increase in 5-HT content by 5-HTP,
        143
    monoamine content, 142-146
    relation to behavior, 144-146, 149-
        152
    stimulation of, 138, 140
    STN neurones, 139, 141
    unit activity of, 136-144
Succinate
    as mitochondrial substrate, 156, 158-
        162
    as redox substrate, 156, 158-162
Synapse (see also Synaptosome)
    concentration of glycoproteins, 210
    postsynaptic membrane, 205
    presynaptic membrane, 205
    relation to learning, memory and
        behavior, 14, 113, 204, 210
    relation to vision, 86

synaptic cleft, 210
Synaptic sites, 16-19, 86
Synaptosome (see also Nerve ending)
  aminoglycolipid content, 204
  GABA binding, 16, 18, 19
  glycine binding, 16, 18, 19
  glycoprotein content, 204
  morphology in brain regions, 35-37
  of isolated and aggregated mice, 16-19
  prostaglandins, 183
  protein content, 16, 17
  protein synthesis, 34, 35
  tryptophan hydroxylase, 223, 224,
    229, 245
  uptake of 5-HT, 37, 38, 224
  uptake of tryptophan, 223, 224, 228-245

T

Tanzi, E., 15
Target-attaching globulins (TAG), 212,
  213, 215-217
Target reagents
  and immunochemical specificity, 212,
    215
Tay-Sachs disease
  and brain glycoproteins, 204, 211
  and glycoprotein 10B, 211
  antibody, 211
  relation to intraneuronal recognition,
    204
Telemetry
  use in completely-free animals, 7-8
Thin-layer chromatography
  for separation of nucleoside alcohols,
    91
Thymidine kinase
  method of determination, 64
  peak activity of seventh post-natal day
    in rat brain, 81
Thyroid hormone (Thyroxine)
  hypothyroid rat, 66-73
  radiothyroidectomy, 71
Thyroxine (see Thyroid hormone)
Time-lapse photography
  method, 135
  of cat, 148, 149
  of circling movements, 148
Tissue diagnostic test
  serum TAG test, 212-217
Tissue and organ specificity, 212

Tolerance
  conventional, 242
  relation to lithium, 242, 243
Transplantation rejection, 212
Tranylcypromine
  effect on brain 5-HT, 115
  use in study of 5-HT, turnover, 28, 29
Tremor
  associated with increased cerebellar
    cyclic-GMP, 184, 189, 191-193
  by injection of 5-HT into substantia
    nigra, 150
  by injection of 5-HTP into substantia
    nigra, 150
  by Nembutal, 149, 150
  cerebellar tremor, 184, 189
  induction by harmaline, 184, 189
  postural type, enhanced by harmaline,
    150
  protection against cerebellar tremor by
    prostaglandin E_2, 189, 191-193
  relation to olivocerebellar system, 189,
    193
Tryptophan
  assay, 227
  hydroxylase, 223-226, 229, 231-245
  increase in brain with hepatic coma, 52,
    53
  induction of hepatic encephalopathy,
    53
  non-albumin bound tryptophan, 53
  uptake, 223-225, 228-230, 233-236,
    238, 239, 242-244
Tryptophan hydroxylase
  assay, 227-228
  effect of
    d-amphetamine, 233, 235, 237
    cocaine, 236, 238-242, 245, 246
    lithium on, 229, 231-236, 239,
      242-245
    morphine, 238
  in nerve endings, 223, 225, 228, 229,
    231-245
Tumor
  astrocytin, 211-220
  content of protein-bound hexose, 211
  distance factor, 210
  DNA replication, 210, 211
  ependymoma-glioma, 210
  genetic (mitotic) activity, 211
  glial cells of, 216

human brain, 216
malignant, brain, 211-220
malignin, 211-220
mucoid-contact functions, 210-220
of brain, 204, 210-220
role of glycoproteins, 204, 210-220
role of recognins, 211-220
serum TAG test, 212-217
Tyrosine
    as precursor for dopamine and nora-
        drenaline, 3
    content of caudate nucleus, 12, 13
    ^{3}H-tyrosine use with dialytrode, 2
    uptake, 228

U

Unit activity
    during rhythmic waves, 136-142, 151,
        152
    of Rh, 136-142, 152
    of STN, 136-142, 152
    of VTA, 136-142, 152
Uptake (see also Low- and High-affinity)
    cocaine, inhibition of tryptophan up-
        take, 236
    high-affinity of 5-HT, 224
    high-affinity of tryptophan, 223
    of 5-HT, 239
    of L-1-^{14}C tryptophan, 228, 230
    of tryptophan, 223, 228-230, 234-236,
        238, 242
Uric acid
    effect on
        age-related differences in protein
            sythesis, 31-34
        age-related differences in retrograde
            amnesia, 30, 31
        ECS-induced change in brain 5-HT,
            28-30
        butyric acid-induced retrograde
            amnesia, 56, 57

cellular 5-HT uptake, 42-46
cellular protein synthesis, 46-49
retrograde amnesia, 26-35, 37, 38,
    42-49, 56-59
synaptosomal 5-HT uptake, 37, 38
synaptosomal protein synthesis,
    34, 35
interactions with
    butyric acid on brain 5-HT content
        and protein synthesis, 57, 58
    5-HT with regard to
        cerebral protein synthesis, 26,
            31-35, 46-49
        changes in 5-HT content and
            disposition, 26, 28-30, 37,
            38, 42-46
        ECS-induced retrograde amnesia,
            26, 28-31, 37, 38, 42-46,
            57-59

V

Valine
    content of caudate nucleus, 12, 13
Ventral tegmental area (VTA)
    and rhythmic slow waves, 133, 135-152
    circling movements, 148, 150, 151
    connection with STN, 144-147
    direct injection of 5-HTP, 144
    direct injection of harmaline, 144, 145,
        150
    effect of
        5-HTP, 143-148
        Nembutal, 135, 143, 145, 148
        raphé lesions, 151
    monoamine content, 142-146
    relation of sleep, 151
Visual system
    anatomical regions, 86
    as a model for sensory deprivation, 86
    polysomes and messenger-RNA, 85-111